An Introduction to the Science

Series in Astronomy and Astrophysics

Series Editors: **M Elvis**, Harvard–Smithsonian Center for Astrophysics
A Natta, Osservatorio di Arcetri, Florence

The Series in Astronomy and Astrophysics includes books on all aspects of theoretical and experimental astronomy and astrophysics. Books in the series range in level from textbooks and handbooks to more advanced expositions of current research.

Other books in the series

The Origin and Evolution of the Solar System
M M Woolfson

Observational Astrophysics
R E White (ed)

Stellar Astrophysics
R J Tayler (ed)

Dust and Chemistry in Astronomy
T J Millar and D A Williams (ed)

The Physics of the Interstellar Medium
J E Dyson and D A Williams

Forthcoming titles

Dust in the Galactic Environment, 2nd edition
D C B Whittet

Very High Energy Gamma Ray Astronomy
T Weekes

Series in Astronomy and Astrophysics

An Introduction to the Science of Cosmology

Derek Raine

Department of Physics and Astronomy
University of Leicester, UK

Ted Thomas

Department of Physics and Astronomy
University of Leicester, UK

Institute of Physics Publishing
Bristol and Philadelphia

British Library Cataloguing-in-Publication Data

A catalogue record for this book is available from the British Library.

ISBN 0 7503 0405 7

Library of Congress Cataloging-in-Publication Data are available

Series Editors: **M Elvis**, Harvard–Smithsonian Center for Astrophysics
A Natta, Osservatorio di Arcetri, Florence

Commissioning Editor: John Navas
Production Editor: Simon Laurenson
Production Control: Sarah Plenty
Cover Design: Victoria Le Billon
Marketing Executive: Laura Serratrice

Published by Institute of Physics Publishing, wholly owned by The Institute of Physics, London

Institute of Physics Publishing, Dirac House, Temple Back, Bristol BS1 6BE, UK

US Office: Institute of Physics Publishing, The Public Ledger Building, Suite 1035, 150 South Independence Mall West, Philadelphia, PA 19106, USA

Typeset in the UK by Text 2 Text, Torquay, Devon
Printed in the UK by J W Arrowsmith Ltd, Bristol

Contents

Preface   **xi**

1   Reconstructing time   **1**
   1.1   The patterns of the stars   1
   1.2   Structural relics   2
   1.3   Material relics   4
   1.4   Ethereal relics   5
   1.5   Cosmological principles   6
   1.6   Theories   7
   1.7   Problems   9

2   Expansion   **10**
   2.1   The redshift   10
   2.2   The expanding Universe   11
   2.3   The distance scale   14
   2.4   The Hubble constant   15
   2.5   The deceleration parameter   16
   2.6   The age of the Universe   16
   2.7   The steady-state theory   17
   2.8   The evolving Universe   18
   2.9   Problems   19

3   Matter   **21**
   3.1   The mean mass density of the Universe   21
       3.1.1   The critical density   21
       3.1.2   The density parameter   21
       3.1.3   Contributions to the density   22
   3.2   Determining the matter density   23
   3.3   The mean luminosity density   24
       3.3.1   Comoving volume   24
       3.3.2   Luminosity function   25
       3.3.3   Luminosity density   25
   3.4   The mass-to-luminosity ratios of galaxies   25
       3.4.1   Rotation curves   26

	3.4.2 Elliptical galaxies	28
3.5	The virial theorem	28
3.6	The mass-to-luminosity ratios of rich clusters	28
	3.6.1 Virial masses of clusters	29
3.7	Baryonic matter	30
3.8	Intracluster gas	31
3.9	The gravitational lensing method	32
3.10	The intercluster medium	33
3.11	The non-baryonic dark matter	33
3.12	Dark matter candidates	34
	3.12.1 Massive neutrinos?	34
	3.12.2 Axions?	35
	3.12.3 Neutralinos?	36
3.13	The search for WIMPS	36
3.14	Antimatter	38
3.15	Appendix. Derivation of the virial theorem	39
3.16	Problems	39

4 Radiation — **41**
4.1	Sources of background radiation	41
	4.1.1 The radio background	41
	4.1.2 Infrared background	43
	4.1.3 Optical background	43
	4.1.4 Other backgrounds	44
4.2	The microwave background	45
	4.2.1 Isotropy	45
4.3	The hot big bang	47
	4.3.1 The cosmic radiation background in the steady-state theory	48
4.4	Radiation and expansion	49
	4.4.1 Redshift and expansion	49
	4.4.2 Evolution of the Planck spectrum	50
	4.4.3 Evolution of energy density	51
	4.4.4 Entropy of radiation	52
4.5	Nevertheless it moves	53
	4.5.1 Measurements of motion	54
4.6	The x-ray background	56
4.7	Problems	58

5 Relativity — **60**
5.1	Introduction	60
5.2	Space geometry	61
5.3	Relativistic geometry	62
	5.3.1 The principle of equivalence	62
	5.3.2 Physical relativity	63
5.4	Isotropic and homogeneous geometry	65

5.4.1	Homogeneity of the 2-sphere	66
5.4.2	Homogeneity of the metric	67
5.4.3	Uniqueness of the space metric	67
5.4.4	Uniqueness of the spacetime metric	68
5.5	Other forms of the metric	68
5.5.1	A radial coordinate related to area	69
5.5.2	A radial coordinate related to proper distance	69
5.6	Open and closed spaces	70
5.7	Fundamental (or comoving) observers	70
5.8	Redshift	71
5.9	The velocity–distance law	73
5.10	Time dilation	74
5.11	The field equations	74
5.11.1	Equations of state	75
5.11.2	The cosmological constant	75
5.11.3	The critical density	76
5.12	The dust Universe	78
5.12.1	Evolution of the density parameter	79
5.12.2	Evolution of the Hubble parameter	79
5.13	The relationship between redshift and time	80
5.13.1	Newtonian interpretation	81
5.14	Explicit solutions	82
5.14.1	$p = 0, k = 0, \Lambda = 0$, the Einstein–de Sitter model	82
5.14.2	The case $p = 0, k = +1, \Lambda = 0$	84
5.14.3	The case $p = 0, k = -1, \Lambda = 0$	86
5.15	Models with a cosmological constant	87
5.15.1	Negative Λ	87
5.15.2	Positive Λ	88
5.15.3	Positive Λ and critical density	88
5.15.4	The case $\Lambda > 0, k = +1$	89
5.16	The radiation Universe	90
5.16.1	The relation between temperature and time	91
5.17	Light propagation in an expanding Universe	92
5.18	The Hubble sphere	93
5.19	The particle horizon	95
5.20	Alternative equations of state	96
5.21	Problems	97
6	**Models**	**101**
6.1	The classical tests	101
6.2	The Mattig relation	102
6.2.1	The case $p = 0, \Lambda = 0$	103
6.2.2	The general case $p = 0, \Lambda \neq 0$	104
6.3	The angular diameter–redshift test	104

	6.3.1	Theory	104
	6.3.2	Observations	106
6.4	The apparent magnitude–redshift test		107
	6.4.1	Theory	107
	6.4.2	The K-correction	108
	6.4.3	Magnitude versus redshift: observations	110
6.5	The geometry of number counts: theory		113
	6.5.1	Number counts: observations	114
	6.5.2	The galaxy number-magnitude test	115
6.6	The timescale test		118
	6.6.1	The ages of the oldest stars	118
6.7	The lensed quasar test		119
6.8	Problems with big-bang cosmology		120
	6.8.1	The horizon problem	120
	6.8.2	The flatness problem	121
	6.8.3	The age problem	122
	6.8.4	The singularity problem	122
6.9	Alternative cosmologies		123
6.10	Problems		124

7   Hot big bang   **128**

7.1	Introduction		128
7.2	Equilibrium thermodynamics		130
	7.2.1	Evolution of temperature: relativistic particles	132
	7.2.2	Evolution of temperature: non-relativistic particles	132
7.3	The plasma Universe		134
7.4	The matter era		135
7.5	The radiation era		136
	7.5.1	Temperature and time	136
	7.5.2	Timescales: the Gamow criterion	137
7.6	The era of equilibrium		138
7.7	The GUT era: baryogenesis		138
	7.7.1	The strong interaction era	139
	7.7.2	The weak interaction era: neutrinos	140
	7.7.3	Entropy and $e^- - e^+$ pair annihilation	140
7.8	Photon-to-baryon ratio		141
7.9	Nucleosynthesis		142
	7.9.1	Weak interactions: neutron freeze-out	143
	7.9.2	Helium	144
	7.9.3	Light elements	146
	7.9.4	Abundances and cosmology	146
7.10	The plasma era		148
	7.10.1	Thomson scattering	148
	7.10.2	Free–free absorption	149

	7.10.3	Compton scattering	150
7.11	Decoupling		151
7.12	Recombination		151
7.13	Last scattering		153
7.14	Perturbations		153
7.15	Appendix A. Thermal distributions		154
	7.15.1	Chemical potentials	154
	7.15.2	Photon energy density	156
	7.15.3	Photon number density	157
	7.15.4	Relativistic neutrinos	157
	7.15.5	Relativistic electrons	158
	7.15.6	Entropy densities	158
7.16	Appendix B. The Saha equation		159
7.17	Appendix C. Constancy of η		159
7.18	Problems		160

8 Inflation   **163**

8.1	The horizon problem		164
8.2	The flatness problem		165
8.3	Origin of structure		165
8.4	Mechanisms		167
	8.4.1	Equation of motion for the inflaton field	168
	8.4.2	Equation of state	169
	8.4.3	Slow roll	170
8.5	Fluctuations		172
8.6	Starting inflation		172
8.7	Stopping inflation		173
	8.7.1	Particle physics and inflation	175
8.8	Topological defects		176
8.9	Problems		176

9 Structure   **179**

9.1	The problem of structure		179
9.2	Observations		180
	9.2.1	The edge of the Universe	181
9.3	Surveys and catalogues		181
9.4	Large-scale structures		182
9.5	Correlations		183
	9.5.1	Correlation functions	183
	9.5.2	Linear distribution	185
	9.5.3	The angular correlation function	185
	9.5.4	Results	185
9.6	Bias		187
9.7	Growth of perturbations		187
	9.7.1	Static background, zero pressure	188

9.7.2	Expanding background	189
9.8	The Jeans' mass	190
9.9	Adiabatic perturbations	192
9.10	Isocurvature (isothermal) perturbations	193
9.11	Superhorizon size perturbations	194
9.12	Dissipation	194
9.13	The spectrum of fluctuations	194
9.14	Structure formation in baryonic models	196
9.15	Dark matter models	197
9.15.1	Growth of fluctuations in dark matter models	197
9.16	Observations of the microwave background	198
9.17	Appendix A	200
9.18	Appendix B	202
9.19	Problems	203

10 Epilogue **205**
10.1	Homogeneous anisotropy	205
10.1.1	Kasner solution	206
10.2	Growing modes	207
10.3	The rotating Universe	208
10.4	The arrow of time	208

Reference material **210**
	Constants	210
	Useful quantities	210
	Formulae	211
	Symbols	212

References **213**

Index **217**

Preface

In this book we have attempted to present cosmology to undergraduate students of physics without assuming a background in astrophysics. We have aimed at a level between introductory texts and advanced monographs. Students who want to know about cosmology without a detailed understanding are well served by the popular literature. Graduate students and researchers are equally well served by some excellent monographs, some of which are referred to in the text. In setting our sights somewhere between the two we have aimed to provide as much insight as possible into contemporary cosmology for students with a background in physics, and hence to provide a bridge to the graduate literature. Chapters 1 to 4 are introductory. Chapter 7 gives the main results of the hot big-bang theory. These could provide a shorter course on the standard theory, although we would recommend including part of chapter 5, and also the later sections of chapter 6 on the problems of the standard theory, and some of chapter 8, where we introduce the current best buy approach to a resolution of these problems, the inflation model. Chapters 5 and 6 offer an introduction to relativistic cosmology and to the classical observational tests. This material does not assume any prior knowledge of relativity: we provide the minimum background as required. Chapters 1 to 4 and some of 5 and 6 would provide a short course in relativistic cosmology. Most of chapter 5 is a necessary prerequisite for an understanding of the inflationary model in chapter 8. In chapter 9 we discuss the problem of the origin of structure and the correspondingly more detailed tests of relativistic models. Chapter 10 introduces some general issues raised by expansion and isotropy. We are grateful to our referees for suggesting improvements in the content and presentation.

We set out to write this book with the intention that it should be an updated edition of *The Isotropic Universe* published by one of us in 1984. However, as we began to discard larger and larger quantities of the original material it became obvious that to update the earlier work appropriately required a change in the structure and viewpoint as well as the content. This is reflected in the change of title, which is itself an indication of how far the subject has progressed. Indeed, it would illuminate the present research paradigm better to speak of the *Anisotropic Universe*, since it is now the minor departures from exact isotropy that we expect to use in order to test the details of current theories. The change of title is at least in part a blessing: while we have met many people who think of the 'expanding

Universe' as the Universe, only more exciting, we have not come across anyone who feels similarly towards the 'isotropic Universe'.

We have also taken the opportunity to rewrite the basic material in order to appeal to the changed audience that is now the typical undergraduate student of physics. So no longer do we assume a working knowledge of Fourier transforms, partial differentiation, tensor notation or a desire to explore the tangential material of the foundations of the general theory of relativity. In a sense this is counter to the tenor of the subject, which has progressed by assimilation of new ideas from condensed matter and particle physics that are even more esoteric and mathematical than those we are discarding. Consequently, these are ideas we can only touch on, and we have had to be content to quote results in various places as signposts to further study.

Nevertheless, our aim has been to provide as much insight as possible into contemporary cosmology for students with a background in physics. A word of explanation about our approach to the astrophysical background might be helpful. Rather than include detours to explain astrophysical terms we have tried to make them as self-explanatory as required for our purposes from the context in which they appear. To take one example. The reader will not find a definition of an elliptical galaxy but, from the context in which the term is first used, it should be obvious that it describes a morphological class of some sort, which distinguishes these from other types of galaxy. That is all the reader needs to know about this aspect of astrophysics when we come to determinations of mass density later in the book.

A final hurdle for some students will be the mathematics content. To help we have provided some problems, often with hints for solutions. We have tried to avoid where possible constructions of the form 'using equations ... it is readily seen that'. Nevertheless, although the mathematics in this book is not in itself difficult, putting it together is not straightforward. You will need to work at it. As you do so we have the following mission for you.

It is sometimes argued, even by at least one Nobel Laureate, that cosmologists should be directed away from their pursuit of grandiose self-titillation at the taxpayers' expense to more useful endeavours (which is usually intended to mean biology or engineering). You cannot counter this argument by reporting the contents of popular articles—this is where the uninformed views come from in the first place. Instead, as you work through the technical details of this book, take a moment to stand back and marvel at the fact that you, a more or less modest student of physics, can use these tools to begin to grasp for yourself a vision of the birth of a whole Universe. And in those times of dark plagues and enmities, remember that vision, and let it be known.

D J Raine
E G Thomas

Chapter 1

Reconstructing time

1.1   The patterns of the stars

It is difficult to resist the temptation to organize the brightest stars into patterns in the night sky.   Of course, the traditional patterns of various cultures are entirely different, and few of them have any cosmological significance. Most of the patterns visible to the naked eye are mere accidents of superposition, their description and mythology representing nothing more than Man's desire to organize his observations while they are yet incomplete. . . .

It is the task of scientific cosmology to construct the history of the Universe by organizing the relics that we observe today into a pattern of evolution. The first objective is therefore to identify all the relics. There has been substantial progress in the last decade, but this task is incomplete. We do not yet know all of the material constituents of the Universe, nor do we have a full picture of their structural organization. The second task is to find a theory within which we can organize these relics into a sequence in time. Here, too, there has been substantial progress in developing a physics of the early Universe, but our current understanding is perhaps best described as schematic.

Suppose though that we knew both the present structure of the Universe and the relevant physics. Unfortunately there are at least two reasons why we could not simply use the theory to reconstruct history by evolving the observed relics backwards in time. The first is thermodynamic. Evolution from the past to the future involves dissipative processes which irreversibly destroy information. We therefore have to guess a starting point and run the system forward in time in the hope of ending up with something like the actual Universe. This is not easy, and it is made more difficult by the second problem, which involves the nature of the guessed starting point. The early Universe, before about 10^{-10} s, involves conditions of matter that are qualitatively different from experience and which can only be investigated theoretically. But, since they involve material conditions unavailable in particle accelerators, these theories can apparently be tested only by their cosmological predictions! This would not matter were it not for the relative dearth of observations in cosmology (i.e. things to predict) relative to the number of plausible scenarios. In consequence the new cosmology is a programme of work in progress, even if progress at present seems relatively rapid.

1.2   Structural relics

In the scientific study of cosmology we are not interested primarily in individual objects, but in the statistics of classes of objects. From this point of view, more important than the few thousand brightest stars visible to the naked eye are the statistics written in the band of stars of the 'Milky Way'. Studies of the distribution and motion of these stars reveal that we are situated towards the edge of a rotating disc of some 10^{11} stars, which we call the Galaxy (with a capital G, or, sometimes, the Milky Way Galaxy). This disc is about 3×10^{20} m in diameter, or, in the traditional unit of distance in astronomy, about 30 kpc (kiloparsecs, see the *List of Constants*, p 210, for the exact value of the parsec).

The division of the stars of the Galaxy into chemical and kinematic substructures suggests a complex history. The young, metal-rich stars (referred to as Population I), with ages $< 5 \times 10^8$ years and low velocity, trace out a spiral structure in the disc. But the bulk of the stellar mass (about 70%) belongs to the old disc population with solar metal abundances, intermediate peculiar velocities and intermediate ages. The oldest stars, constituting Population II, have higher velocities and form a spheroidal distribution. Their metal abundances (which for stellar astrophysicists means abundances of elements other than hydrogen and helium) are as low as 1% of solar abundances.

The agglomeration of stars into galaxies is itself a structural relic. Think of the stars as point particles moving in their mutual gravitational fields interchanging energy and momentum. Occasionally a star will approach the edge of the galaxy with more than the escape velocity and will be lost to the system. Eventually, most of the stars will be lost in this way proving that galaxies are transient structures, relics from a not-too-distant past.

Some 200 globular clusters are distributed around the Galaxy with approximate spherically symmetry. These are dense spherical associations of 10^5–10^7 stars of Population II within a radius of 10–20 pc, which move in elliptical orbits about the galactic nucleus. The Andromeda Nebula (M31), which at a distance of 725 kpc is the furthest object visible to the naked eye, provides us with an approximate view of how our own Galaxy must look to an astronomer in Andromeda. The collection of our near-neighbour galaxies is called the Local Group, Andromeda and the Milky Way being the dominant members out of some 30 galaxies. Galaxies come in a range of types. Some (including our Galaxy and M31) having prominent spiral arms are the *spiral galaxies*; others regular in shape but lacking spiral arms are the *elliptical galaxies*; still others, like the nearby Magellanic Clouds, are classed as *irregulars*.

Imagine now that we turn up the contrast of the night sky so that more distant sources become visible. Astronomers describe the apparent brightness of objects on a dimensionless scale of 'apparent magnitudes' m. The definition of apparent magnitude is given in the *List of Formulae*, p 211; for the present all we need to know is that fainter objects have numerically larger magnitudes (so you need larger telescopes to see them). As systems with apparent magnitude in the visible waveband, m_v, brighter than $m_v = +13$ can be seen, we should be able to pick out a band of light across the sky in the direction of Virgo. This contains the Virgo cluster of galaxies at the centre of which, at a distance of about 20 Mpc, is the giant elliptical galaxy M87. Virgo is a rich irregular cluster of some 2000 galaxies. Most of this band of light comes from other clusters of various numbers of galaxies of which our Local Group is a somewhat inconspicuous and peripheral example. The whole collection of galaxies makes up the Virgo Supercluster (or Local Supercluster). The Local Supercluster is flattened, but, like the elliptical galaxies and in contrast to the spirals, the flattening is not due to rotation. Several other large structures can be seen (about 20 structures have been revealed by detailed analysis) with a number concentrated towards the plane of the Local Supercluster (the Supergalactic plane). Even so these large structures are relatively rare and the picture also reveals a degree of uniformity of the distribution of bright clusters on the sky.

Turning up the contrast still further, until we can see down to an apparent magnitude of about 18.5, the overall uniformity of the distribution of individual galaxies becomes apparent. This can be made more striking by looking only at the distribution of the brightest radio sources. These constitute only a small fraction of bright galaxies and provide a sparse sampling of the Universe to large distances. The distribution appears remarkably uniform, from which we deduce that the distribution of matter on the largest scales is isotropic (the same in all directions) about us.

So far we have considered the projected distribution of light on the sky. Even here the eye picks out from the overall uniformity hints of linear structures, but it is difficult to know if this is anything more than the tendency of the human brain to form patterns in the dark. The advent of an increasing number of large telescopes has enabled the distribution in depth to be mapped as well. (This is achieved by measurements of redshifts, from which distances can be obtained, as will be explained in chapter 2.) The distribution in depth reveals true linear structures. Some of these point away from us and have been named, somewhat inappropriately, the 'Fingers of God'. They appear to place the Milky Way at the centre of a radial alignment of galaxies, but in fact they result from our random motion through the uniform background, rather like snow seen from a moving vehicle. (There is a similar well-known effect in the motion of the stars in the Hyades cluster.) Of course, this leaves open the question of what causes our motion relative to the average rest frame of these distant galaxies. It may be the gravitational effect of an enhanced density of galaxies (called the Great Attractor, see section 4.5). These three-dimensional surveys also indicate large voids of up to 50 Mpc in diameter containing less than 1% of the number of galaxies that would be expected on the basis of uniformity. Nevertheless, as we shall see in detail in chapter 9, these large-scale structures, both density enhancements and voids, are relatively rare and do not contradict a picture of a tendency towards overall uniformity on a large enough scale.

1.3   Material relics

With the luminous matter of each of the structures we describe we can associate a mass density. The average density of visible matter in the Galaxy is about 2×10^{-21} kg m^{-3}, obtained by dividing the total mass by the volume of the disc. The Local Group has a mean density of 0.5×10^{-25} kg m^{-3}. The average density of a rich cluster, on the other hand, is approximately 2×10^{-24} kg m^{-3}, while that of a typical supercluster may be 2×10^{-26} kg m^{-3}. The average density clearly depends on both the size and location of the region being averaged over. The result for rich clusters goes against the trend, but these contain only about 10% of galaxies. Then the dominant trend is towards a decrease in mass density the larger the sample volume. What is the limit of this trend?

It is simplest to assume that the process reaches a finite limit, beyond which point larger samples give a constant mass density. This would mean that, on some scale, the Universe is uniform. But in principle the density might oscillate with non-decreasing amplitude or the density might tend to zero. Both of these possibilities have been considered, although neither of them very widely, and the former not very seriously. The latter is called a hierarchical Universe. We can arrange for it as follows. Take clusters of order n to be clustered to form clusters of order $n + 1$ ($n = 1, 2, \ldots$). The clusters of order $n + 1$, within a cluster of order $n + 2$, are taken to be separated by a distance much larger than $n^{-1/3}$ times the separation of clusters of order n within a cluster of order $n + 1$. In such a system the concept of an average density is either meaningless, or useless, since the density depends on the volume of space averaged over, except in the infinite limit, when the fact that it is zero tells us very little.

One might think that there is nothing much to say about a third possibility— the uniform (homogeneous) Universe. This is not the case. Suppose that the Universe consists of randomly arranged clusters of some particular order m which are themselves therefore not clustered. (Of course, the random arrangement can produce fluctuations, accidental groupings; by random, and not clustered, we mean clumped no more than would be expected on average by chance.) On a scale larger than the mth cluster this Universe is homogeneous and has a finite mean density. Alternatively, one might contemplate an arrangement in which some, but not all mth-order clusters are clustered, but the rest are randomly distributed. This too would be, on average, uniform. In fact, the Universe is homogeneous on sufficiently large scales, but neither of these arrangements quite matches reality. We shall return to this question in chapter 9.

Just as the clustering of matter into stars tells us something about the history of the galaxy, so the clustering on larger scales carries information about the early Universe. But matter has more than its mass distribution to offer as a relic. There is its composition too. In nuclear equilibrium the predominant nuclear species is iron (or in neutron-poor environments, nickel), because iron has the highest binding energy per nucleon. In sufficiently massive stars, where nuclear equilibrium is achieved, the result is an iron or nickel core. From the fact that 93% of the nuclei in the Universe, by number, are hydrogen and most of the remaining 7% are helium, we can deduce that the Universe can never have been hot enough for long enough to drive nuclear reactions to equilibrium. The elements heavier than helium are, for the most part, not primordial. On the other hand, helium itself is, and the prediction of its cosmic abundance, along with that of deuterium and lithium, is one of the achievements of big-bang cosmology.

A surprising fact about the matter content of the Universe is that it is not half antimatter. At high temperatures the two are interconvertible and it would be reasonable to assume the early Universe contained equal quantities of each which were later segregated. The observational evidence, and the lack of a plausibly efficient segregation mechanism, argue against this. Either a slight baryon excess was part of the initial design or the laws of physics are not symmetric between matter and antimatter. The latter is plausible in a time-asymmetric environment (section 7.7). We shall see that the absence of other even more exotic relics than antimatter is a powerful constraint on the physics of the early Universe (chapter 8).

On the other hand, there must be some exotic relics. The visible matter in the Universe is insufficient to explain the motion under gravity of the stars in galaxies and the galaxies in rich clusters. Either gravity theory is wrong or there exists matter that is not visible, dark matter. Most cosmologists prefer the latter. This in itself is not surprising. After all, we live on a lump of dark matter. However, the dark matter we seek must (almost certainly) be non-baryonic (i.e. not made out of protons and neutrons). This hypothetical matter could be known particles (for example neutrinos if these have a mass) or as yet undiscovered particles. Searches have so far revealed nothing. Thus the most important material relic remains to be uncovered.

1.4   Ethereal relics

The beginning of physical cosmology can be dated to the discovery of the cosmic background radiation, the fossilized heat of the big bang. This universal microwave radiation field carries many messages from the past. That it is now known to have an exact blackbody spectrum has short-circuited many attempts to undermine the big-bang orthodoxy (see, for example, section 4.3.1). Since the radiation is not in equilibrium with matter now, the Universe must have been hot and dense at earlier times to bring about thermal equilibrium (as we explain in chapters 4 and 7). Equilibrium prevailed in the past; the radiation must have cooled as a result of the uniform expansion of space. This ties in with the increasing redshift of more distant matter which, through the theory of relativity, links the redshift to expansion.

The cosmic background radiation also provides a universal rest frame against which the motion of the Earth can be measured. (This does not contradict relativity, which states only that empty space does not distinguish a state of rest.) That the overall speed of the Earth, around 600 km s^{-1}, turns out to be unexpectedly large is also probably an interesting relic in itself. Once the effect of the Earth's motion has been subtracted, the radiation is found to be the same in all directions (isotropic) to an extent much greater than anticipated. Thus, when they interacted in the past, the inhomogeneities of matter were impressed on the radiation to a lesser degree than expected from the current fluctuations in density. This points to an additional component to the matter content, which would allow inhomogeneities to grow more rapidly. Fluctuations in the matter density could therefore evolve to their present values from smaller beginnings at the time when the radiation and matter interacted. Since this extra component of matter is not seen, it must be dark. Thus the cosmic background radiation provides evidence not only for the isotropy of the Universe, and for its homogeneity, but for the existence of dark matter as well.

We shall find that background radiation at other wavelengths is less revealing. For the most part it appears that the extragalactic radio background is integrated emission from discrete sources and so too, to a large extent, is the background in the x-ray band. In principle, these yield some cosmological information, but not readily, and not as much as the respective resolved sources, the distributions of which again confirm isotropy and homogeneity at some level. The absence now of a significant background at optical wavelengths (i.e. that the sky is darker than the stars) points to a finite past for the Universe. The sky was not always dark; the observation that it is so now implies an origin to time (section 2.8).

1.5   Cosmological principles

The *cosmological principle* states that on large spatial scales the distribution of matter in the Universe is homogeneous. This means that the density, averaged over a suitably large volume, has essentially the same, non-zero value everywhere at the present time. The cosmological principle was originally introduced by Einstein in 1917, before anything was known about the large-scale distribution of matter beyond our Galaxy. His motivation was one of mathematical simplicity. Today the principle is more securely based on observation. We know it does not hold on small scales where, as we have stated, matter exhibits a clear tendency to cluster. Sheets or wall-like associations of galaxies and regions relatively empty of galaxies, or voids, ranging in size up to around 100 Mpc have been detected (1 Mpc, or Megaparsec, equals 10^6 pc). However, a transition to homogeneity is believed to occur on scales between 100 and 1000 Mpc, which is large compared to a cluster of galaxies, but small compared to the size of the visible Universe of around 9000 Mpc. In any case, we can adopt the cosmological principle as a working hypothesis, subject to observational disproof.

A key assumption of a different kind is the *Copernican principle*. This states that, for the purpose of physical cosmology at least, we do not inhabit a special location in the Universe. The intention is to assert that the physical laws we can discover on Earth should apply throughout the Universe. There is some evidence for the consistency of this assumption. For example, the relative wavelengths and intensities of spectral lines at the time of emission from distant quasars are consistent with exactly the atomic physics we observe in the laboratory. On the other hand, it is difficult to know what would constitute incontrovertible evidence against it. In any case, most, if not all, cosmologists would agree that there is nothing to be gained by rejecting the Copernican principle.

We can bring the cosmological and Copernican principles together in the following way. The distribution of galaxies across the sky is found to be isotropic on large angular scales (chapter 9). According to the Copernican principle, the distribution will also appear isotropic from all other locations in the Universe. From this it can be shown to follow that the distribution is spatially homogeneous (chapter 2), hence that the cosmological principle holds.

Thus, the cosmological principle is a plausible deduction from the observed isotropy. Nevertheless, it is legitimate to question its validity. For example, while it might apply to the visible Universe, on even much larger scales we might find that we are part of an inhomogeneous system. This is the view taken in inflationary models (chapter 8). Alternatively, although apparently less likely, the tendency of matter to cluster could extend beyond the visible Universe to all length scales. In this case it would not be possible to define a mean density and the cosmological principle would not be valid. (The matter distribution would have a fractal structure.) Galaxy surveys which map a sufficiently large volume of the visible Universe should resolve this issue. Two such surveys are being carried out at the present time. Currently the evidence is in favour of the cosmological principle despite a few dissenting voices.

The cosmological principle is also taken to be valid at all epochs. Evidence in support of this comes from the cosmic microwave background which is isotropic to about one part in 10^5 (chapter 4). This implies that the Universe was very smooth when it was 10^5 years old (chapter 9) and also that the expansion since that time has been isotropic to the same accuracy.

1.6 Theories

The general theory of relativity describes the motion of a system of gravitating bodies. So too does Newtonian gravity, but at a lower level of completeness. For example, Newtonian physics does not include the effects of gravity on light. The greater completeness of relativity has been crucial in providing links between the observed relics and the past. We shall see that relativity relates the homogeneity and isotropy of a mass distribution to a redshift. (Strictly, to a universal shift of spectral lines either to the red or to the blue.) Historically, homogeneity and isotropy were theoretical impositions on a then blatantly clumpy 'Universe' (the nearby galaxies) and the universal redshift had to await observational revelation. But relativity, if true, provided the framework in which uniformly distributed matter implies universal redshifts and the expansion of space. Within this framework the cosmic background radiation gave us a picture of the Universe as an expanding system of interacting matter and radiation that followed, up to a point, laboratory physics. The picture depends on the truth of relativity, but not on the details of the theory, only on its general structure which, nowadays, is unquestioned.

Nevertheless, laboratory physics can take us only so far back in time: effectively 'the first three minutes' starts at around 10^{-10} s, not at zero. This does not matter if one is willing to make certain assumptions about conditions at 10^{-10} s. Most of the discussion in this book will assume such a willingness on the part of the reader, not least because it is a prerequisite to understanding the nature of the problems. In any case, if the Universe is in thermal equilibrium at 10^{-5} s, much of the preceding detail is erased.

However, the limitations of laboratory physics do matter if one wishes to investigate (or even explain) the assumptions. Exploration of the very early Universe, before 10^{-10} s, depends on the extrapolation of physical laws to high energy. This extrapolation is tantamount to a fundamental theory of matter. That such a theory might be something fundamentally new is foreshadowed by the problems that emerge in the hot big-bang model once the starting assumptions are queried (chapter 6).

We do not have a fundamental theory of matter, so we have to turn the problem round. What sort of characteristics must such theories have if they are to leave the observed relics of the big bang (and not others)? To ask such a question is to turn from theories to scenarios. The scenario that characterizes the new cosmology is analogous to (or perhaps even really) a change of phase of the material content.

In the Universe at normal temperatures we do not (even now) observe the superconducting phase of matter. Only if we explore low temperatures does matter exhibit a transition to this phase. In the laboratory such temperatures can be investigated either experimentally or, if we are in possession of a theory and the tools to work out its consequences, theoretically. The current scenarios for the early Universe, which we shall look at in chapter 8, are analogous, except that here we explore changes of state at extremely high temperatures (about 10^{28} K perhaps). A feature of these scenarios is that they involve a period of exponentially rapid expansion of the Universe, called inflation, during the change of phase. In this picture the visible Universe is a small part, even at 10^{10} light years, of a finite system, apparently uniform because of its smallness.

This picture has had some notable successes, but it remains a programme. It has one major obstacle. This is *not* that we do not possess a theory of matter. It is that the theory of gravity is not yet complete! The theory of relativity is valid classically, but it does not incorporate the effect of gravity into quantum physics or of quantum physics into gravity. (Which of these is the correct order depends on where you think the fundamental changes are needed.) Of course, one can turn this round. Just as the very early Universe has become a test-bed for theories of matter, so the first moments may become a test of quantum gravity: the correct theory must predict the existence of time.

1.7 Problems

Problem 1. *Given that the Universe is about 10^{10} years old, estimate the size of the part of it visible to us in principle ('the visible Universe'). Assuming that the Sun is a typical star, use the data in the text and in the list of constants together with the mean density in visible matter, $\rho \sim 10^{-28}$ kg m^{-3} to get a rough estimate of the number of galaxies in the visible Universe. (This number is usually quoted as about 10^{11} galaxies, the same as the number of stars in a galaxy.)*

Problem 2. *Estimate the escape velocity from the Galaxy. Estimate the lifetime of a spherical galaxy assuming the stars to have a Maxwell–Boltzmann distribution of speeds. (You can use the approximation $\int_y^\infty x^2 \mathrm{e}^{-x^2}\,\mathrm{d}x \sim \frac{y}{2}\mathrm{e}^{-y^2}$ as $y \to \infty$.)*

Problem 3. *The speed of the Galaxy relative to the Virgo cluster is around a few hundred km s^{-1}. Deduce that the Supercluster is not flattened by rotation.*

Problem 4. *The wavelength λ of a spectral line depends on the ratio of electron mass to nuclear mass through the reduced electron mass such that $\lambda \propto 1 + m/M$. Explain why measurements of the ratio of the wavelength of a line of Mg^+ to that of a line of H can be used to determine whether the ratio of electron to proton mass is changing in time. With what accuracies do the line wavelengths need to be measured to rule out a 1% change in this mass ratio? (Pagel 1977)*

Problem 5. *Relativistic quantum gravity must involve Newton's constant G, Planck's constant h and the speed of light c. Use a dimensional argument to construct an expression for the time at which, looking back into the past, quantum gravity effects must become important. (This is known as the Planck time.) What are the corresponding Planck length and Planck energy? What are the orders of magnitude of these quantities?*

Chapter 2

Expansion

2.1   The redshift

The wavelengths of the spectral lines we observe from an individual star in the Galaxy do not correspond exactly to the wavelengths of those same lines in the laboratory. The lines are shifted systematically to the red or to the blue by an amount that depends on the velocity of the observed star relative to the Earth. The overall relative velocity is the sum of the rotational velocity of the Earth, the velocity of the Earth round the Sun and the Solar System around the centre of the Galaxy, in addition to any velocity of the star. The rotational velocity of the Galaxy makes the largest contribution to the sum, so it gives the order of magnitude of this relative velocity. At the radial distance of the Solar System it is about 220 km s^{-1}. The first-order Doppler shift in wavelength, $\Delta\lambda = \lambda_e - \lambda_o$, for a velocity $v \ll c$, is given by

$$\frac{\Delta\lambda}{\lambda_e} = \frac{v}{c}, \tag{2.1}$$

where λ_o and λ_e are the observed and emitted wavelengths. The redshift, z, is *defined* by

$$z = \frac{\Delta\lambda}{\lambda_e}, \tag{2.2}$$

so the ratio of observed-to-emitted wavelength is

$$\frac{\lambda_o}{\lambda_e} = 1 + z. \tag{2.3}$$

Negative z corresponds, of course, to a blue shift. For the Ca II line at 3969 Å, for example, we expect a shift of order 4 Å corresponding to a redshift $z \approx \pm 10^{-3}$. This is much larger than the width of the spectral lines due to the motion of the emitting atoms in a stellar photosphere (see problem 6).

If we look at an external galaxy we see not the individual stars but the integrated light of many stars. The spectral lines will therefore be broadened by $\Delta\lambda/\lambda_e \sim 10^{-3}$ because of the motion of the stars in that galaxy. For members of the Local Group, typical velocities are of the order of a few hundred km s^{-1}, and therefore give rise to red or blue shifts of the same order as the line broadening. For some of the brightest nearby galaxies we find velocities ranging from 70 km s^{-1} towards us to 2600 km s^{-1} away from us. But as we go to fainter and more distant galaxies, the shifts due to the local velocities become negligible compared with a systematic redshift. If, in this preliminary discussion, we interpret these redshifts as arising from the Doppler effect, then the picture we get is of a concerted recession of the galaxies away from us.

Radio observations of external galaxies in the 21 cm line of atomic hydrogen give velocities which agree with optical values over the range -300 km s^{-1} to $+4000$ km s^{-1}, as expected for the Doppler effect. Nevertheless, one should be aware that the naive interpretation of the redshift as a straightforward Doppler effect, convenient as it is for an initial orientation, is by no means an accurate picture. We will justify the use of the Doppler formula $v = cz$ for small redshifts z in section 5.8.

2.2   The expanding Universe

In view of the isotropy of the matter distribution about us, which we discussed in chapter 1, it is natural to assume that the observed recession of galaxies follows the same pattern in whatever direction we look. In fact, the isotropy of the matter distribution is logically distinct from the isotropy of the expansion. Nevertheless it is difficult to imagine how a direction in which galaxies were receding more slowly could avoid having an excess of brighter, nearer galaxies, unless a non-isotropic expansion had contrived to produce an isotropic matter distribution just at our epoch. This latter possibility would be too much of a cosmic conspiracy, so the expansion about us is assumed to be isotropic. We shall show later that this picture is supported by observational evidence. Invoking the Copernican principle, we conclude that observers everywhere at the present time see an isotropic expansion. The overall effect of expansion is therefore a change of length scale. This implies that the Universe remains homogeneous as it expands, in conformity with the cosmological principle.

A standard way to visualize such a centreless expansion is to imagine the cooking of a uniform currant loaf of unlimited size. As the loaf cooks the spacing between currants increases, so any given currant sees other currants receding from it. The loaf always looks similar: only its length scale changes. The homogeneously expanding Universe is analogous to the loaf with the currants replaced by galaxies and the dough replaced by space. Notice though that the loaf expands homogeneously even if the currants are not distributed uniformly through it, which would not be true in the Universe, where the galaxies affect the expansion rate.

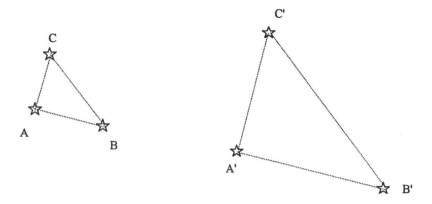

Figure 2.1. Three galaxies at A, B and C at time t_i expand away from each other to A′, B′ and C′ at time t.

We arrive at the important conclusion that the expansion of the Universe can be described by a single function of time, $R(t)$. This function is called the *scale factor*. We will now give a more formal proof of this, which will enable us to derive the relationship between the recessional velocity of a galaxy and its distance from us.

Consider the three widely separated galaxies, A, B and C, shown in figure 2.1 at an initial time t_i and at a later time t, and look at the expansion from the point of view of the observer in A.

Isotropy implies that the increase in distance AB → A′B′ be the same as AC → A′C′; but, from the point of view of the observer at C, isotropy requires that the expansion in AC be the same as in BC. The sides of the triangle are expanded by the same factor. Adding a galaxy out of the plane to extend the argument to three dimensions, we see that isotropy about each point implies that the expansion is controlled by a single function of time $R(t)$, from which the ratios of corresponding lengths at different times can be obtained. If AB has length l_i at time t_i, then at time t its length is

$$l(t) = l_i \frac{R(t)}{R(t_i)}. \tag{2.4}$$

Differentiating this expression with respect to time gives the relative velocity, $v(t) = dl(t)/dt$, of B and A:

$$v(t) = \frac{l_i}{R(t_i)} \frac{dR(t)}{dt} = \frac{\dot{R}(t)}{R(t)} l(t), \tag{2.5}$$

where $\dot{R} = dR/dt$ and we have used equation (2.4) to substitute for $l_i/R(t_i)$ in terms of values at a general time t.

In this picture there is no edge to the distribution of galaxies—such an edge would violate the cosmological principle—so the expansion should not be thought of as an expansion of galaxies into an empty space beyond. (There is nothing outside the whole system for it to expand into.) It is better thought of as an expansion of the intervening space between galaxies. The velocity in equation (2.5) is an expansion velocity: it is the rate at which the intervening space between A and B is increasing and carrying them apart. (The process is analogous to the way in which the currants in our currant loaf are carried apart by the expanding dough.) It is important to realize that equation (2.5) is true for all separations l. We shall refer to equation (2.5) as the *velocity–distance law* after Harrison (2000). For l sufficiently large, expansion velocities can exceed the speed of light. We shall return to this point in section 5.18 where we shall show that there is no conflict with the special theory of relativity.

For small velocities we can use the non-relativistic Doppler formula to get $v = cz$ (see problem 7) which, on substitution into (2.5), gives the following linear relation between redshift and distance:

$$cz(t) = l(t)\frac{\dot{R}}{R}. \tag{2.6}$$

Therefore the assumption of a homogeneous and isotropic expansion leads to a linear relation between the redshift of a galaxy and its distance from us, at least for sufficiently nearby galaxies. This prediction provides us with a test of the expanding space picture.

In order to make a meaningful comparison with observation, it is necessary to decide what exactly it is in our not exactly uniform Universe that is supposed to be expanding according to (2.6). Such objects as atoms, the Earth, the Sun and the Galaxy do not expand, because they are held together as bound systems by internal electrical or gravitational forces. Consider, for example, the Galaxy, mass M_G, radius r_G. The gravitational potential (GM_G/r_G) is a rough measure of the strength of the internal binding. (It is not exact because the Galaxy is not spherical!) It corresponds to a dimensionless escape velocity, $v_{esc}/c = (2GM_G/r_Gc^2)^{1/2}$ of order $v_{esc}/c \sim 10^{-3}$. The recessional velocity of the edge of the Galaxy as seen from its centre would be given by (2.5) as $v/c = (\dot{R}/R)r_G/c$. For $\dot{R}/R = 10^2$ km s^{-1} Mpc^{-1} (see section 2.4), this is $v/c \sim 3 \times 10^{-6}$, which is negligible compared to the escape velocity. So the internal gravitational force dominates over the expansion of the Universe. A typical rich cluster has a mass 10^2–10^3 times a galactic mass, and a radius 200 times that of the Galaxy, and this again leads to a bound system. For nearest-neighbour clusters however, taking the intercluster distance to be about three cluster diameters, we obtain $v_{esc}/c \sim 5 \times 10^{-4}$, whereas $(\dot{R}/R)r/c \sim 2 \times 10^{-3}$. Thus separate clusters are typically not bound together by gravity. Consequently, to a first approximation, we should regard the clusters of galaxies, rather than the galaxies themselves, as the basic units, or 'point particles', of an expanding Universe.

2.3   The distance scale

To investigate the validity of equation (2.6) we need to find the relationship between redshift and distance for galaxies which are far enough from us to be participating in the universal Hubble flow, but not so far away that the relationship (2.6) does not apply. In practice this means getting the distances and redshifts of galaxies which lie beyond the Virgo Supercluster, of which we are an outlying member, out to a redshift not exceeding $z \approx 0.2$ (Sandage 1988).

In order to obtain the distance to a galaxy we climb out along the rungs of a distance ladder. All but one of these rungs can involve relative distances, but one at least must be absolute. Nowadays the absolute measurement is provided by the diameter of the Earth's orbit about the Sun, which can be found by radar ranging. The distances to nearby stars can then be obtained by measuring their angular shift against the background of more distant stars when they are viewed from opposite ends of the Earth's orbit. This parallax method can be used out to a distance of about 50 pc. The Hipparcos satellite has been used to obtain distances to about 120 000 stars by this means.

The luminosities of these stars can be found from their known distances by using the inverse square law. This gives rise to the notion of a *standard candle*. The luminosity, or equivalently the absolute magnitude, of a star of given spectral type is known by reference to these nearby stars. From the measurement of its apparent magnitude, the distance of a star, which lies beyond the range of the parallax method, can be determined from the inverse square law. Stars of a type that can be identified in this way are standard candles. The brightest stars and, in particular, variable stars can be used as standard candles. Cepheid variable stars are particularly useful standard candles: their pulsation periods are related to their luminosities in a known way, and the long period ones are intrinsically very bright so can be seen at large distances. The Hubble Space Telescope can record the light curves of Cepheid variable stars in galaxies out to a distance of about 20 Mpc, a volume of space that encompasses the Virgo cluster and contains of the order of 10^3 galaxies. The light curves of the Cepheids give the distance to their host galaxy and calibrate the luminosity of the host. If necessary the brightest galaxies themselves can then be used as standard candles. Also, the luminosity of supernovae of type Ia (SNe Ia) occurring in these galaxies can be obtained from Cepheid distances. Recent research has shown that SNe Ia are, in fact, better standard candles than the galaxies. However, they are relatively rare events in a given galaxy, which limits their usefulness.

In his pioneering investigation Hubble (1929) found a 'roughly linear' relation between redshift and distance. Later studies carried out to higher redshifts have confirmed the linear relation

$$cz = H_0 l \qquad (2.7)$$

which is known as Hubble's law. Figure 2.2 shows a recent plot of velocity $v = cz$ against distance using supernovae of type Ia as distance indicators (see Filippenko

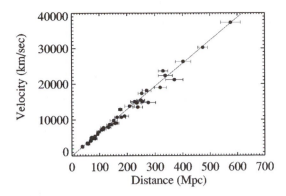

Figure 2.2. A plot of velocity against distance obtained from observations of supernovae (from Turner and Tyson 1999).

and Riess 1998). Note the linearity of the plot in figure 2.2, which is in accord with the prediction of equation (2.6). So the expanding space model passes this first test.

Once H_0 is known, a measurement of redshift alone can be used, together with Hubble's law, to obtain the distance to a galaxy, provided that the galaxy is far enough away for the contribution of random velocities to the redshift to be unimportant. Note that Hubble's law, in the form of equation (2.7), can be used only out to $z \approx 0.2$. Beyond $z \approx 0.2$ the effect of the expansion on the propagation of light is no longer negligible and the interpretation of the redshift as a local Doppler shift is no longer viable. A specific cosmological model is then needed in order to turn a redshift into a distance. We will return to this topic when we discuss cosmological models in chapter 6.

2.4 The Hubble constant

The quantity H_0 appearing in equation (2.7) is the Hubble constant. Comparing equation (2.5) with equation (2.7) leads us to define the Hubble parameter

$$H = \frac{\dot{R}(t)}{R(t)}.$$

The Hubble parameter H is a function of time but is independent of position at any time. The subscript 0 denotes the value at the present time, t_0, so $H_0 = H(t_0)$.

The value of the scale factor at any time depends on the arbitrary choice of length scale l_0 so it cannot be a measurable quantity. The Hubble parameter, on the other hand, is a *ratio* of scale factors, so is independent of choice of scale. It is therefore an important quantity in cosmology: it is an observable measure of the rate at which the Universe is expanding.

To obtain the value of the Hubble constant from observation is, in principle, straightforward but, in practice, it is fraught with difficulty. It is obtained from the slope of a plot of redshift z against the distance of galaxies out to $z \sim 0.2$. Redshifts can be measured accurately, but obtaining accurate distances to galaxies is much more difficult. Hubble's original estimate was $H_0 = 550$ km s^{-1} Mpc^{-1}. Since Hubble carried out his work in the 1930s the discovery of systematic errors in his distance measurements has brought H_0 down to a value between 50 and 100 km s^{-1} Mpc^{-1}. Because of this uncertainty in the exact value of H_0 it is usual to write

$$H_0 = 100h \text{ km s}^{-1} \text{ Mpc}^{-1}$$

and to keep track of the factor h in all formulae so that they can be adjusted as required. Recent progress in distance measurement has reduced the uncertainty. At the time of writing there is a growing consensus that $H_0 = 67 \pm 10$ km s^{-1} Mpc^{-1} (see Filippenko and Riess 1998). Note that the slope of figure 2.2 gives $H_0 = 64$ km s^{-1} Mpc^{-1}. We shall henceforth take $H_0 = 65$ km s^{-1} Mpc^{-1}, or $h = 0.65$, for numerical evaluations.

2.5   The deceleration parameter

Since the Hubble parameter may change with time we introduce another parameter that gives its rate of change. The simplest suggestions might be to take $\ddot{R}$ or $\dot{H}$. However, the former would have a value that depends on both the units of time and on the arbitrary absolute value of the length scale, so would not be measurable, while the latter depends on the choice of units for time, which is regarded as inconvenient. We therefore choose the dimensionless number

$$q = -\frac{\ddot{R}R}{\dot{R}^2}$$

as a measure of the deceleration of the expansion of the Universe. The quantity q is called the *deceleration parameter*. The minus sign is included because it was thought that the expansion should be slowing down under the mutual gravitational attraction of matter as the Universe gets older. This would mean that q_0, the present value of q, would conveniently be positive. Recent observations cast some doubt on this, and there is growing evidence that the sign of q_0 is negative.

2.6   The age of the Universe

If we assume for the moment that the velocity of expansion is constant in time, then two galaxies separated by a distance d_0 move apart with a velocity $v = Hd = H_0 d_0$. The distance separating this pair of galaxies (and any other pair) was zero at a time H_0^{-1} before the present. The quantity H_0^{-1} is therefore the present age of a Universe undergoing constant expansion and, consequently,

H_0^{-1} provides an estimate of the age of the actual Universe. If the deceleration parameter is positive then the Hubble parameter was greater in the past, so the age of the Universe is somewhat less than H_0^{-1}. Note that, from the cosmological principle, no one point of space can be regarded as being the centre of the expansion, so all points must have been coincident (in some sense) at the initial time.

Hubble's original estimate for H_0^{-1} gives $H_0^{-1} \approx 2 \times 10^9$ years. It was clear at the time that this was too low since it is shorter than the then known age of the Earth. What was not clear then was whether this represented a problem with the observations or a failure of the theory, thus opening the door to alternative models of the cosmos.

2.7 The steady-state theory

The evolution of astrophysical systems within the Universe, in particular the irreversible processing of material by stars, implies that they have a finite lifetime. It then becomes difficult to see how the Universe could be infinitely old. The expanding Universe of finite age, the big-bang model which we have previously assumed, represents one way out of the problem. There is, however, another possibility, a Universe in which new matter is being continuously created. The rate of creation is too small to be directly observed in the laboratory and there is no other empirical foundation for this hypothesis. Nevertheless the idea can be investigated to see where it leads. The simplest assumption we can make is that the large-scale properties of the Universe do not change with time. In that case matter must be created at the rate needed to maintain the mean mass density at its present value (see problem 11) and H must be constant and equal to its present value H_0. Consequently

$$\mathrm{d}l/\mathrm{d}t = H_0 l,$$

which can be integrated to give

$$l = l_0 \exp(H_0 t).$$

The steady-state expansion is therefore exponential, with the scale factor

$$R(t) = R(0) \exp(H_0 t),$$

and not linear as one might have guessed! The exponential curve is self-similar in the sense that increasing t by a given amount is equivalent to rescaling l_0. There is therefore no privileged origin for time and the result is indeed compatible with the notion of a steady state.

The steady-state theory can be subjected to observational tests both of its basic non-evolutionary philosophy and in its detailed prediction of constant H. It fails both these types of test: it fails to account for the evolution of radio sources (Wall 1994), x-ray sources (Boyle *et al* 1994) and optical galaxies (Dressler *et al*

1987) and a constant H is ruled out by the shape of the redshift–distance relation at large redshifts. Perhaps the most serious difficulty is its failure to account in any natural way for the blackbody spectrum of the microwave background radiation (chapter 5). It also fails to account for the fact that about 25% by mass of the baryonic matter in the Universe is in the form of helium, because this is too much to be attributed to nucleosynthesis in stars. The steady-state theory is of historical interest, because it is a properly worked out example of an alternative to the big-bang theory and demonstrates the sort of tests that such alternatives must survive if they are to be serious competitors. However, nowadays the research programme of cosmology is the same as that of any other branch of physics: to explore the known laws of physics to their limits.

2.8   The evolving Universe

The most surprising aspect of the night sky, once one has absorbed the presence of the stars, is the existence of the dark spaces between the stars. The paradoxical nature of the darkness, first pointed out by Kepler, Halley and Le Chésaux, has come to be known as *Olbers' paradox*. To state the problem assume the Universe to be uniformly filled with a number density n of stars, each having a luminosity L. If the solid angle occupied by these stars were to cover the whole celestial sphere the average surface brightness of the night sky would be the same as that of the average star. This condition would be satisfied if the Universe were to extend to 6×10^{37} m or more (problem 12). The light from the most distant stars would take 7×10^{21} years to reach us. But stars do not live so long. In other words, the darkness of the night sky is witness to the evolution of the stars. The expansion of the Universe redshifts the light from distant galaxies and so complicates this argument. But, except for cases like the steady-state picture, where the redshift is the only effect, the complications do not alter the discussion significantly.

If we were able to look back into the Universe to a redshift of about 300 we should see no galaxies, because at this time the galaxies would have overlapped. Galaxies, and the large-scale structures associated with them, have formed since then (chapter 9). Once created, the galaxies themselves also evolve with time. One can point to the distribution of quasars (galaxies with non-stellar emission from their nuclei) which, from the number counts, appears to have peaked around $z = 2$. And the number of galaxies as a function of luminosity has changed, with more luminous galaxies having been more common in the past.

The key to understanding the evolution of the Universe is the microwave background radiation (chapter 4). The Universe was not only more compressed in the past but also hotter. The resulting picture is the hot big-bang model. To understand this, and develop its consequences, we must study the evolution of matter and radiation in an interacting system (chapter 7).

This will lead us to certain features of the Universe which are apparently not accounted for by evolution (chapter 6). In the hot big bang these have to be accounted for by postulating certain initial conditions. In fact, the epithet 'big bang' was originally coined as a term of ridicule for a theory that produces the Universe, partially formed, out of a singularity. It has become clear in recent years that the big-bang theory is incomplete, and that there may be ways of making the early evolution more convincing (chapter 8).

Not only is our evolution from the past of interest, but so too is the course of our future evolution. The question of whether the Universe will go on expanding forever or will eventually halt and collapse has always been a central issue in cosmology. On the basis of current theory we can make certain predictions (chapter 5) depending on the equation of state of matter in the Universe (chapter 3). And finally (chapter 10), what of the isotropy? Is the present symmetry a product of early evolution from an anisotropic beginning that will evolve away or is it a principle that will remain forever?

2.9 Problems

Problem 6. *Show that the Ca II line at 3969 Å emitted from the photosphere of stars in the Galaxy would be expected to exhibit a wavelength shift of order 4 Å, corresponding to a redshift $z \approx \pm 10^{-3}$, as a result of galactic rotation. Show that this is much larger than the width of the spectral lines due to the motion of the emitting atoms in the stellar photosphere.*

Problem 7. *Show that the redshift can be defined in terms of frequencies rather than wavelengths by*

$$z = \frac{v_e - v_o}{v_o}$$

and that for small z this becomes $z = |\delta v|/v_e$.

The relativisitic Doppler shift of a source receding with speed v is

$$v_o = v_e \left(\frac{1 - v/c}{1 + v/c} \right)^{1/2}.$$

Deduce that the Doppler formula applied to a body receding slowly would give $v = cz$.

Problem 8. *The Hipparcos satellite was able to measure parallaxes to an accuracy of 10^{-3} arc sec. What error does this give for a star at 50 pc?*

Problem 9. *The Hubble Space Telescope (HST) was used to observe Cepheids in the Virgo cluster with periods P of between 16 and 38 days. The period–luminosity relation for Cepheids in the Large Magellanic Cloud at 55 kpc, from HST data, is*

$$M_v = -2.76 \log P - 1.4,$$

where M_V is the absolute visual magnitude. If the fit to the Virgo Cepheids gives an apparent visual magnitude

$$m_V = -2.76 \log P + 29.89$$

what is the distance to Virgo?

Problem 10. *In a Universe with scale factor $R(t) \propto t^p$ (p a positive constant, but not necessarily an integer) show that the deceleration parameter q is a constant and $q \lessgtr 0$ according to whether $p \gtrless 1$. Compare the exact ages of these Universes with the estimates using the approximate relation: age $\sim H_0^{-1}$. Is this estimate ever exact?*

Problem 11. *Show that in order to maintain a constant density of (say) 10^{-26} kg m^{-3} in the steady-state Universe, mass must be created at a rate of about $10^{-43}h$ kg m^{-3} s^{-1}. One version of the theory has the matter created as neutrons. What additional level of β-radioactivity would one expect to find in the Earth if this were the case?*

Problem 12 (Olbers' paradox) *Show that the night sky would have a surface brightness equal to that of a typical star if stars lived for in excess of 7×10^{21} years. Take the stars to be distributed uniformly through the Universe.*

Chapter 3

Matter

3.1   The mean mass density of the Universe

The ultimate fate of the Universe is determined, through its gravity, by the amount and nature of the matter it contains. The amount of mass in the world is therefore a quantity of considerable importance in cosmology. One of the most unexpected results of modern cosmology is that most of this mass is not only unseen, but neither is it made from the protons, neutrons and electrons of normal matter.

3.1.1   The critical density

It is the usual practice to express the mean mass density ρ in terms of a *critical density* ρ_c defined by

$$\rho_c = \frac{3H^2}{8\pi G},$$ (3.1)

where H is the Hubble parameter. The critical density is a *function of time* through H. For the simplest cosmological models a Universe with a density equal to or less than the critical value expands forever, while a Universe with more than the critical density is destined to collapse (chapter 5).

3.1.2   The density parameter

The ratio

$$\Omega = \rho/\rho_c$$ (3.2)

is referred to as the *density parameter* and is commonly used as a convenient dimensionless measure of density.

Substituting $H_0 = 100h$ km s^{-1} Mpc^{-3} into equation (3.1) gives

$$(\rho_c)_0 = 1.88 \times 10^{-26} h^2 \text{ kg m}^{-3},$$ (3.3)

where, as usual, the subscript 0 denotes the value of a quantity at the present time. To rough order of magnitude, this is the observed density. Thus we need to be able to measure the density to better than an order of magnitude in order to determine the fate of the Universe. This was the original motivation behind accurate density measurements, although, as we shall see, the motivation has expanded.

Expressing the critical density in cosmological units gives

$$(\rho_c)_0 = 2.76 \times 10^{11} h^2 M_\odot \text{ Mpc}^{-3} \qquad (3.4)$$

which is about one galactic mass per cubic megaparsec. From (3.2) and (3.3) the current density is

$$\rho_0 = 1.88 \times 10^{-26} h^2 \Omega_0 \text{ kg m}^{-3}. \qquad (3.5)$$

3.1.3   Contributions to the density

The total density Ω includes contributions from matter Ω_m, and from radiation Ω_r. In addition we shall see in chapter 8 that, because of the zero point energy of a quantum field, the cosmological 'vacuum' need not be the same as the state of zero energy; so Ω may also contain a contribution from the mass equivalent of the energy of the 'vacuum'. This is labelled Ω_λ. Thus

$$\Omega = \Omega_m + \Omega_r + \Omega_\lambda. \qquad (3.6)$$

If there are any other sources of mass-energy they will also contribute to Ω and must be added to (3.6). The corresponding quantities at the present time are $(\Omega_m)_0$, etc. In order not to overencumber the notation, we shall define

$$\Omega_M = (\Omega_m)_0; \qquad \Omega_R = (\Omega_r)_0; \qquad \Omega_\Lambda = (\Omega_\lambda)_0,$$

so, we have also

$$\Omega_0 = \Omega_M + \Omega_R + \Omega_\Lambda.$$

A precise determination of the present total density Ω_0 is expected from measurements of the anisotropy of the cosmic microwave background radiation on small angular scales to be made by the Planck and MAP satellites. At the present time anisotropy measurements appear to be forcing Ω_0 towards $\Omega_0 = 1$ (Balbi *et al* 2000). This topic will be treated in chapter 9. The traditional route to Ω_0 is via separate determinations of Ω_M, Ω_R and Ω_Λ and any other contributions. Agreement between the two approaches will provide a check on the consistency of our cosmology.

The contribution of Ω_R to the total density parameter Ω_0 at the present time is small (problem 13), but it was the dominant contributor to Ω in the early Universe. The evidence for a vacuum energy will be discussed in chapter 8. In this chapter we will consider the component Ω_M which includes stellar matter and any other form of matter that can cluster under its own gravity.

3.2   Determining the matter density

There are two approaches to the determination of mass. It can either reveal itself through its gravity or through the radiation, if any, it emits. We use the former to obtain, for example, the mass of the Sun from the centripetal acceleration of the Earth. In this example, the method actually gives the total mass within the Earth's orbit, including any non-luminous matter, but the orbits of the other planets show that the measured mass is concentrated within the Sun. This approach can be extended to obtain the masses of galaxies from stellar motions, and the masses of binary galaxies and of clusters of galaxies from galaxy motions. Ultimately, one obtains the mass (and hence the density) of the Universe from the motions of distant galaxies.

One might think that knowing the masses of stars one could obtain the masses of galaxies by counting stars. There are two problems. First, we cannot be sure of counting all the faint stars. Second, there may be material in a galaxy in a form other than stars. These problems can be overcome if we can determine the average amount of matter (including dark matter) associated with a given light output, and if we can measure the average total light output (including that from sources too faint to be identified individually). Thus, the classical method for obtaining the mean mass density, ρ_M, due to matter employs the second of these approaches through the relationship

$$\rho_M = \ell \frac{M}{L}, \tag{3.7}$$

where ℓ is the mean luminosity per unit volume of matter and M/L the mean ratio of mass to luminosity for a representative sample of the matter.

More precisely, the universal luminosity density ℓ due to galaxies is the light radiated into space in a given waveband per second per unit volume (section 3.3). This is a fairly well determined quantity and has a value

$$\ell = (2 \pm 0.2) \times 10^8 h L_\odot \, \text{Mpc}^{-3} \tag{3.8}$$

for blue light (Fukugita *et al* 1998). The corresponding quantity M/L is the value of the global mass-to-luminosity ratio (or, equivalently, mass-to-light ratio), also measured in the blue band.

The mass-to-light ratios of rich clusters of galaxies are used to determine ρ_M. These systems are taken to be large enough to be a representative sample of matter, so their mass-to-light ratios can be assumed to approach the global value for the Universe as a whole. Note that since the mean mass of a cluster is obtained from its gravity, this method of determining ρ_M includes all forms of matter, whether luminous or dark, that has an enhanced density in a cluster. This excludes matter which is too hot to cluster appreciably (because random motions exceed escape velocity), and also any other smooth component of mass (but not the x-ray emitting gas in clusters). The mean density within a cluster is of the order of a hundred times the global mean mass density. So any uniformly distributed mass makes an insignificant contribution to the gravity of a cluster, since its density is not enhanced within the cluster.

Substituting the values quoted earlier for ρ_c and l into equation (3.7) gives the mass-to-light ratio corresponding to a critical density in matter:

$$(M/L)_{\text{crit.}} = \frac{2.76 \times 10^{11} h^2 M_\odot \text{ Mpc}^{-3}}{2.0 \pm 0.2 \times 10^8 h L_\odot \text{ Mpc}^{-3}} \tag{3.9}$$
$$= (1390 \pm 140) h (M_\odot / L_\odot).$$

Thus

$$\Omega_M = \frac{M/L}{(M/L)_{\text{crit.}}}. \tag{3.10}$$

A method for obtaining Ω_M via gravity uses the Hubble plot together with the assumption that $\Omega_0 = 1$ (see chapter 6). In principle, the amount of gravitating matter can be obtained from the curvature of the plot of redshift versus distance at large redshift. This automatically measures the global value.

In section 3.3 we outline the determination of the mean luminosity density. Section 3.4 describes the determination of the mass-to-luminosity ratio of a galaxy. From this we conclude that the visible part of a galaxy is immersed inside a large halo of dark matter which contains most of its mass. In the appendix, at the end of the chapter, we derive the *virial theorem* relating the gravitational potential energy and internal kinetic energy of a gravitating system in equilibrium. This provides a tool for the determination of the mass-to-luminosity ratio of galaxy clusters, which is outlined in section 3.6. Finally everything is brought together and a value for the density parameter is obtained. The conclusion that most of the matter is non-luminous or dark matter is explained. A further and very surprising conclusion is that most of the matter content of the Universe is not protons, neutrons and electrons but an as yet unidentified non-baryonic type of matter. A discussion of what this dark matter might be, and experimental attempts to detect it, is covered in sections 3.12 and 3.13.

3.3   The mean luminosity density

3.3.1   Comoving volume

We define a *comoving volume* as a region of space that expands with the Universe. Therefore, if galaxies were neither created nor destroyed, the number of galaxies in a comoving volume would not change with time. Since physical volumes are expanding the number of galaxies per unit volume is decreasing if the number per comoving volume is constant. The use of comoving volumes factors out the expansion so one can see the intrinsic effects of galaxy evolution.

3.3.2   Luminosity function

The luminosity function of the galaxies $\Phi(L)$ is defined so that the number of galaxies per unit comoving volume with luminosity in the range L to $L + dL$ is

$$dN = \Phi(L)dL.$$

The mean luminosity density l is then given by

$$l = \int_0^\infty \Phi(L)L\,dL. \tag{3.11}$$

A reasonable fit to the observed number density of galaxies is given by the Schecter function (Schecter 1976) which has the form

$$\Phi(L)\,dL = \Phi_*(L/L_*)^{-\alpha}e^{-L/L_*}dL/L_*, \tag{3.12}$$

with L_* a constant.

3.3.3   Luminosity density

Integrating equation (3.11) with this function gives

$$l = \Phi_*L_*\Gamma(2+\alpha), \tag{3.13}$$

where Γ is the gamma function (problem 15). Approximate values of the parameters in equation (3.12) are: $\alpha \approx 1.0$, $\Phi_* \approx 10^{-2}h^3$ Mpc^{-3}, $\Gamma(3) = 2$ and $L_* \approx 10^{10}h^{-2}L_\odot$. Quoted values for l fall in the range given in equation (3.8).

3.4   The mass-to-luminosity ratios of galaxies

We shall follow the standard practice of expressing the ratio of mass to luminosity, M/L, in solar units. By definition the mass-to-light ratio of the Sun is $1.0M_\odot/L_\odot$. Stars less massive than the Sun have M/L greater than unity and stars more massive than the Sun have M/L less than unity, because the luminosity grows with mass faster than direct proportionality. The M/L ratio for the typical mixture of stars making up a galaxy is in the range 1–10. So if galaxies were composed of the stars that we see, and nothing else, they would have mass-to-luminosity ratios in this range. In fact when the mass of an elliptical or spiral galaxy is determined through its gravity, and its luminosity determined from the direct measurement of its light output, the M/Ls obtained range up to $100h$ for spirals and perhaps four times higher for ellipticals. Evidently the bulk of the mass of a galaxy is dark.

3.4.1   Rotation curves

Let us consider, first of all, how the mass of a spiral galaxy is determined. A spiral galaxy consists of a circular disc of gas and stars in rotation about its centre. The distribution of mass in the galaxy determines, through Newtonian gravity, the orbital velocities of the stars and gas clouds. The method used to obtain the mass of a spiral galaxy is, in principle, the same as that used to obtain the mass of the Sun. For a body of mass m moving in a circular orbit of radius R with velocity V about the Sun, Newtonian mechanics gives

$$mV^2/R = GMm/R^2.$$

So measurements of R and V enable the mass M of the Sun to be obtained from

$$M = RV^2/G. \tag{3.14}$$

Note that for orbits about a centrally concentrated mass, such as for bodies in the Solar System orbiting the Sun, this equation gives $RV^2 = $ constant. Thus the orbital velocities of the planets fall off with their distance from the Sun according to the Keplerian law $V \propto R^{-1/2}$. In a spiral galaxy, in contrast, the mass is distributed in both the disc and also throughout an extended roughly spherical dark halo the existence of which, as we shall see, is inferred from the measured velocities.

The rotation of a galaxy which is not face on to our line of sight gives different redshifts for stars on opposite sides of the centre. By subtracting the recessional velocity of the galaxy we obtain the stellar velocities as a function of radius R in the rest frame of the galaxy. Finally these velocities are corrected for the effect of geometric projection to give the orbital velocities V in the plane of the disc. A plot of the orbital velocities of stars, or of gas clouds, against radial distance from the centre of a spiral galaxy gives a *rotation curve*. Rotational velocities beyond the visible part of the galactic disc at ∼20 kpc out to a few times this value are obtained by measuring the Doppler shifts in the 21 cm emission from clouds of neutral hydrogen. Figure 3.1 shows a rotation curve together with a plot of the luminosity of the galactic disc against radius.

Beyond the radii reached by using rotation curves, the motion of satellite galaxies can be used to probe the mass distribution out to a distance of the order of 200 kpc (Zaritsky *et al* 1993).

The surprising feature of the rotation curves is that, as shown in figure 3.1, the orbital velocities do not show a Keplerian fall off, that is V is not proportional to $R^{-1/2}$, even when the velocity measurements are extended to radii well beyond the visible part of the galaxy. Rotation curves are typically flat, i.e. $V = $ constant, out to the largest radii to which measurements can be taken. Evidently, the mass is not concentrated in the visible galaxy, but is much more extended.

The flatness of the rotation curves, and the velocities of satellites, can be explained if a galaxy is embedded in a large massive spheroidal halo the density

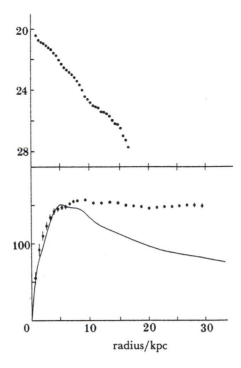

Figure 3.1. Upper panel: A logarithmic plot of surface brightness as a function of radius for the galaxy NGC 3198. Lower panel: The observed rotation curve of the galaxy (points) and the rotation speed calculated from the mass associated with the light profile only (full line) (from Albada and Sancisi 1986).

of which falls off with radius as $\rho \propto R^{-2}$. This implies that the mass of matter $M(R)$, inside radius R, is proportional to R and hence, from (3.14), that the orbital velocity $V = $ constant. In practice more detailed models are used in which the mass distribution is not assumed to be spherical but is itself deduced from the distribution of velocities.

Using these mass determinations, it is found that the mass-to-light ratios, M/L, for spiral galaxies increase with radius, from $\sim 10h$ for the visible part of a galaxy up to $\sim 100h$ at a radius of about 200 kpc (Bahcall *et al* 1995). This means that the halo extends out to at least ~ 10 times the visible radius of a galaxy and contains most of the mass of the galaxy.

It is obvious that these high values of M/L imply that the extended halo material contains matter that is much less luminous than normal stars. It is therefore referred to as *dark matter*. Thus, the majority of matter in spiral galaxies is dark.

Finally, note that we have assumed here that Newtonian gravity is valid on galactic scales. Sometimes the deficit of luminous gravitating matter is interpreted as evidence that gravity follows something other than an inverse square law on these scales (e.g. Milgrom 1986). There appears to be no independent evidence for this view and it is generally discounted.

3.4.2 Elliptical galaxies

For elliptical galaxies the rotation curve cannot be used to obtain the distribution of mass as a function of radius, since ellipticals do not, in general, have a concerted rotational motion: the stellar orbits have random directions like the orbits of stars in globular clusters. The mass of the visible part of an elliptical galaxy can be obtained from the measured velocity dispersion of starlight using the virial theorem (section 3.5), in the same way that the masses of rich galaxy clusters are found (section 3.6). Mass beyond the visible part of the galaxy can be estimated from the motion of satellite galaxies, globular clusters, hot x-ray emitting gas or neutral hydrogen when these are present. The results are that for ellipticals, also, there is evidence for the existence of extended massive halos. The mass-to-light ratios are found to increase with scale and are generally larger, within the same radius, than those for spirals by approximately a factor of three. They probably range up to values $\sim 400h$ (Bahcall *et al* 1995). Thus, most of the matter in elliptical galaxies is dark.

3.5   The virial theorem

The *virial theorem* is a conservation law for systems of interacting particles which have achieved a state of equilibrium. It is therefore of less general applicability than the conservation of energy and momentum, which are valid whether or not a system is in equilibrium. If T is the total kinetic energy and V the total potential energy of the particles, the virial theorem states that

$$2T + V = 0. \tag{3.15}$$

In the appendix (p 39) we give a proof of this result for point particles moving under their mutual gravitational attraction.

The theorem specifies that the sum of twice the kinetic energy plus the potential energy is exactly zero for the limited class of systems to which it applies. In particular, two particles in elliptical orbits about each other do not satisfy the theorem in this form, but a pair of particles in circular orbits do. Clearly some care is needed in its application. We shall be considering large clusters of galaxies. In this case the theorem is applicable provided the clusters have had time to *relax* to equilibrium configurations, i.e. if, on average, the structure is unchanging.

3.6   The mass-to-luminosity ratios of rich clusters

We saw in chapter 1 that galaxies show a tendency to cluster. For example, the Milky Way belongs to the Local Group, a system containing upwards of 30 members. At the other extreme of scale are the rich clusters containing numbers of galaxies ranging from several hundred to over a thousand. These clusters have masses in the range $2 \times 10^{14} M_\odot$ to $1 \times 10^{15} M_\odot$ within a radius of 1–2 Mpc. The Virgo and Coma clusters are examples of rich clusters. There are three ways of obtaining the mass of a cluster: (1) through the motion of the individual galaxies; (2) through measurements of the temperature of the hot intracluster gas; and (3) through the gravitational lensing of background galaxies. The mass-to-light ratios obtained from these three methods are in broad agreement with each other. In this section we explain method (1) and in section 3.8 we explain method (2) and outline method (3) in section 3.9.

3.6.1 Virial masses of clusters

The oldest method of estimating the mass of a cluster uses the measured redshifts of individual member galaxies to deduce the typical velocities of these galaxies in the frame of the cluster. It is then found that the time required for a typical galaxy to cross the cluster is much less than the age of the Universe. It follows that if the cluster were not a bound system there would have been ample time for the galaxies to disperse. In addition, to a first approximation at least, the regular spherical distribution of galaxies throughout a rich cluster, such as Coma, resembles the distribution of atoms in a finite volume of an isothermal sphere of self-gravitating gas. Consequently, we may assume that such systems are in equilibrium and apply the virial theorem in order to determine the mean mass of the cluster.

If the ith galaxy has mass m_i and velocity v_i, the kinetic energy T of the cluster is given by

$$T = \tfrac{1}{2} \sum m_i v_i^2 = \tfrac{1}{2} M \langle v^2 \rangle,$$

where $\langle v^2 \rangle = \sum m_i v_i^2 / \sum m_i$ is the mean square velocity and $M = \sum m_i$ is the total mass of the cluster. Note that $\langle v^2 \rangle$ is the true velocity dispersion: it is three times the measured one since, of the three velocity components of a galaxy, only the component along the line of sight contributes to the redshift.

The potential energy of the cluster, assumed to be spherical, is given by

$$V = -\alpha \frac{GM^2}{R_A},$$

where $3/5 < \alpha < 2$ depending on the distribution of mass in the cluster; its value can be estimated from the surface density of galaxies on the sky. The quantity $R_A = 1.5h^{-1}$ Mpc is a measure of the radius of the cluster called the Abell radius.

Substituting the kinetic and potential energies into the virial theorem, equation (3.15), to find the cluster mass gives

$$M = \frac{R_A \langle v^2 \rangle}{\alpha G}.$$

Using cluster masses obtained in this way gives mass-to-luminosity ratios which lie in the range $\sim 200h$–$400h$. This is of the order of the typical mass-to-light ratio for a mixture of galaxies, depending on the proportions of spirals and ellipticals. We deduce that most of the mass in clusters must be associated in some way with the constituent galaxies. In section 3.8 we shall see that the presence of hot gas in clusters accounts for about 10–15% of the mass of a cluster.

The fact that the dark matter in clusters must be associated mainly with the galaxies implies that no significant additional material is involved in cluster formation. Nevertheless, there is evidence from gravitational lensing that cluster galaxies lose their halos through tidal stripping within the cluster, so this material is now distributed throughout the cluster. (A typical cluster galaxy is less massive than a field galaxy.) This is somewhat different from the way this was presented in older textbooks where the cluster mass is often compared with the *visible* mass in galaxies to deduce the presence of dark matter *somewhere* in the cluster. Similarly, the mass in intracluster gas is often compared to the visible mass in the cluster (see later).

Measurements of mass-to-luminosity ratios on even larger scales are less certain but appear not to rise significantly beyond the values obtained for clusters (Bachall *et al* 1995). If this is the case we can use the cluster M/Ls to estimate Ω_M. Substituting the cluster M/L ratios into equation (3.10) yields

$$\Omega_M \approx 0.15\text{–}0.3. \tag{3.16}$$

Some simulations of the formation of galaxy structure suggest that luminous matter may cluster more strongly than the dark matter. If this were the case there would be a bias to dark matter on large scales which would increase the large-scale mass-to-light ratio and increase the value of Ω_M measured on large scales. A plot of magnitude against redshift, using supernovae of type Ia as standard candles (section 6.4), gives an independent and large-scale value of $\Omega_M = 0.28$. The agreement with (3.16) is evidence that biasing (section 9.6) is not a problem.

3.7 Baryonic matter

The value of Ω_M obtained in the preceding section, equation (3.16), presents us with a surprising problem. To see what this is we anticipate the results of chapter 7 on the synthesis of light elements in the big bang. This theory predicts the amounts of the light elements ^{4}He, ^{2}H, ^{3}He and ^{7}Li produced in the early Universe as a function of the matter density at the time of nucleosynthesis. One might therefore think that a value for Ω_M could be deduced by comparing the predicted yields of these elements with values of their primeval abundances obtained from observation; and, in particular, the higher the value of the matter density at the epoch of nucleosynthesis the smaller the amount of deuterium that is produced. Knowledge of the primeval abundance of deuterium therefore provides an upper bound on the density of matter. The requisite deuterium abundance can now be obtained from spectroscopic observations of highly redshifted clouds of hydrogen gas. Such clouds have metal abundances about 10^{-3} of the solar value and so contain nearly pristine primeval material. At the time of writing, observations of a number of clouds with the Keck telescope give consistent values for the primordial deuterium abundance (Tytler *et al* 1996). Let us write Ω_B ('B' for baryonic matter) for the current value that would be deduced for Ω_M based on these observations (supplemented by observations of helium and lithium abundances). We obtain

$$\Omega_B = (0.02 \pm 0.004)h^{-2} \tag{3.17}$$

(Schramm and Turner 1998). Taking $h = 0.65$ together with the upper limit in (3.17) gives $\Omega_B < 0.06$.

The problem is now clear: however we allow for the uncertainties, the upper limit for the matter density deduced from nucleosynthesis is clearly incompatible with the value deduced from direct observations. It will become clear from chapter 7 why cosmologists are reluctant to relinquish the theory of nucleosynthesis. In any case, a recent attempt to draw up a direct inventory of baryonic matter (Fukugita *et al* 1998) also leads to the conclusion that $\Omega_B \sim 0.02$.

The preferred solution is to conclude that Ω_M in (3.16) is measuring a different quantity from Ω_B in (3.17). The density Ω_M includes all gravitating matter whereas the density Ω_B includes only the component involved in nucleosynthesis, namely baryonic matter (neutrons and protons and the accompanying electrons required for charge neutrality). Comparison of Ω_B with the lower bound on Ω_M given in (3.16) forces us to conclude that the majority of the dark matter in galaxies and clusters cannot be in the form of baryons. We have

$$\Omega_M = \Omega_B + \Omega_D,$$

where Ω_D is the current density due to dark matter. Identifying the nature of the non-baryonic mass density constitutes the dark matter problem. We shall return to this later.

3.8 Intracluster gas

Given that most of the dark matter is non-baryonic one might think that most of the baryons should be in luminous matter. This is not the case. We can estimate the contribution of optically luminous matter, Ω_{LUM}, from the M/L ratios for the visible parts of galaxies. These, as we stated in section 3.4, lie in the range 1–10. So Ω_{LUM} lies in the range 0.001–0.007. Therefore $\Omega_{LUM} \ll \Omega_B$ and we conclude that most of the baryons are not in stars. In fact, within clusters, these baryons are mostly present as hot gas between the galaxies.

X-ray observations of the hot gas between the galaxies in clusters provides us with a second way to estimate Ω_M. The measured surface brightness in x-rays can be used to obtain the density and temperature profiles of the gas and from these the mass of gas within the Abell radius of the cluster can be inferred. It turns out that the mass in gas substantially exceeds the mass in the stars of the constituent galaxies of a cluster and appears to account for all the baryons not present in stars. For example, for the Coma cluster, $M_{gas}/M_{gal} \gg 5h^{-3/2}$ (White *et al* 1993). A composite value for the ratio of the mass in baryons M_B to the total mass M for a cluster is

$$M_B/M = (0.07 \pm 0.007)h^{-3/2} \qquad (3.18)$$

(Evrard 1997, Turner and Tyson 1999). If we assume that this ratio is representative of the global value we can write

$$M_B/M = \Omega_B/\Omega_M. \qquad (3.19)$$

Now substituting this value for M_B/M and the value for Ω_B from (3.17) into equation (3.19) gives

$$\Omega_M = \frac{0.02h^{-2}}{0.07h^{-3/2}} \approx 0.3h^{-1/2}.$$

If, in addition to the gas and visible stars, there were appreciable quantities of baryons locked up in dark bodies such as Jupiter-like objects or brown dwarfs then the ratio M_B/M would be increased and the value of Ω_M reduced. The independent check provided by the determination of Ω_M from the Hubble plot suggests that this is not the case. In addition, searches for gravitational lenses in the galactic halo do not find enough compact objects to account for the dark halo mass.

3.9   The gravitational lensing method

When a light-ray passes through the gravitational field of a large mass it is bent in the same sense as a converging lens. The gravitational field of a spherical mass is a somewhat odd lens in that the amount of bending decreases away from the axis. This means that an extended object on the axis is focused into a ring (an *Einstein* ring) by an intervening spherical mass. If non-spherical objects focus slightly off-axis objects they produce arcs and arclets. Rather amazingly both giant arcs and arclets are observed where background galaxies are focused by intervening clusters. The radius of an arc can be used to determine the central mass concentration in the cluster, and the arclets provide details of its distribution.

At the time of writing the three methods for estimating cluster masses described earlier are in reasonable agreement with each other. Moreover, the resulting value for Ω_M is in good agreement with that obtained from the Hubble plot. Turner (1999) quotes the following value for Ω_M which is based on all methods, assuming $h = 0.65$:

$$\Omega_M \cong 0.35 \pm 0.07. \qquad (3.20)$$

3.10   The intercluster medium

Clusters are formed by the gravitational collapse of regions of enhanced density. In the process they gather up matter, both baryonic and non-baryonic, over a large volume. Consequently we expect the relative proportions to be typical of the Universe as a whole, even though the fraction of luminous matter might be different between clusters and diffuse matter.

We have seen that most of the baryonic matter within clusters is not in stars but in the form of hot gas which is readily detectable through its x-ray emission. Outside rich clusters the baryons in gaseous form are harder to locate because they are in shallower potential wells and therefore cooler. However, surveys of low-energy x-ray emission from small to intermediate size groups of galaxies with the ROSAT satellite detect the presence of associated warm plasma (Mulchaey *et al* 1996). The average plasma mass fraction detected is

$$\frac{M_{\text{plasma}}}{M} = (0.022 \pm 0.005)h^{-3/2}$$

which is lower than the value for rich clusters given in equation (3.18). This difference could arise because the bulk of the plasma is too cool to be detected by this means. The recent detection from the Hubble Space Telescope of faint absorption lines of O VI from this material is relevant, but it is not possible to use this yet to estimate the total mass density of plasma (Tripp *et al* 2000).

We have seen that locally the bulk of baryons are in diffuse matter rather than in stars. This is even more the case at large redshift. Historically, the best known of the early tests for an intergalactic density of neutral hydrogen was the Gunn–Peterson test. This involved an attempt to look for an absorption in Lyα at redshift $z \sim 2$ shifted into the optical spectrum. Since there is no evidence of such absorption by a diffuse component, we can guess the optical depth to be less than, say, $\tau = 0.01$. This implies (Peebles 1993) that the density in smoothly distributed neutral hydrogen is some six orders of magnitude below the local baryon density extrapolated to large redshift. The resolution is partly that the material at these redshifts is clumped, and partly that the smooth component is ionized. The photoionized component can be detected through absorption by He$^+$ ions with the result that this might be a significant component ($\Omega \lesssim 0.006h^{-3/2}$), but may be negligible. The clumped component constitutes the Lyα forest clouds consisting of clumped photoionized material and is the major contributor to the baryon mass. The total baryon mass at this epoch is consistent, within large uncertainties, to that observed in gas and stars at the present time.

3.11   The non-baryonic dark matter

We have seen that there is strong evidence from the dynamics of galaxies and galaxy clusters that the matter which makes up the Universe constitutes about 35% of the critical density. At the same time the theory of nucleosynthesis in the early Universe leads to the conclusion that the only matter that we know of, matter made from baryons, amounts to, at most, 6% of the critical density ($\Omega_B \lesssim 0.06$ compared to $\Omega_M \sim 0.35$). The conclusion is that at least 85% of the matter is of an unknown form. Attempts have been made to escape from this unpalatable admission of our ignorance. For example, considerable effort has gone into attempts to modify the standard picture of big-bang nucleosynthesis by considering the possibility that the baryons were distributed in a clumpy fashion at the epoch of nucleosynthesis (Schramm and Turner 1998). By adjusting the density of the clumps and the typical distance between them it was hoped that the predicted baryon density could be raised sufficiently to close the gap between Ω_B and Ω_M and thus remove the need for non-baryonic dark matter. In the end it turned out not to be possible to do this while retaining the successes of the standard theory. The failure of these attempts has reinforced our confidence in the standard picture of nucleosynthesis.

Another possibility is that the dark matter is baryonic but that it was in a form that could not participate in nucleosynthesis. For example that it was locked up in black holes. This idea also has its difficulties—for example, how to form black holes of appropriate mass?

A further argument for the existence of non-baryonic dark matter is that it is needed to explain the formation of galaxies. The idea is that galaxies formed from the gravitational collapse of some initial density perturbations (for the existence of which there is independent evidence, as we shall see in chapter 9). Computational simulations of galaxy formation cannot produce the large-scale structure that we see today without the presence of a cold and weakly interacting form of matter which readily clumps. At the present time, therefore, the existence of non-baryonic dark matter seems to be an essential ingredient for a self-consistent cosmology.

3.12 Dark matter candidates

The dark matter must consist of massive particles (to be consistent with structure formation) which interact weakly (to be consistent with nucleosynthesis). Such weakly interacting massive particles are now referred to as WIMPS. Moving on to the candidates that have been suggested for the dark matter particles, we find that particle physics provides three front-runners for consideration: massive neutrinos, axions and neutralinos.

3.12.1 Massive neutrinos?

The standard big-bang scenario predicts a background of neutrinos which decoupled from the radiation when the Universe was about 1 s old (section 7.7.2). Their present number density is about 113 cm^{-3} for each of the three neutrino flavours. Thus, the total number density of neutrinos is similar to the number density of background photons, which is 412 cm^{-3}. If they are massless in accordance with the standard model of particle physics then their mass density is negligible at the present time. On the other hand, a single neutrino flavour with a rest mass of about 40 eV/c^2 would give a critical density (problem 63). At present there is evidence that the sum of the masses of the three flavours of neutrinos in the range 0.05 eV $< m_\nu < 8$ eV (Ahmad *et al* 2001). With these limits the temperature at which the neutrinos cease to be relativistic lies in the range $6 \times 10^2 < T < 9 \times 10^4$. We shall assume that neutrinos are still relativistic at the epoch when the density of matter and of radiation were equal (section 5.16) which occurs at a redshift of about 10^4 or a temperature of about 3×10^4 K. This assumption is obviously not valid for a small range of neutrino masses at the upper end of that allowed by experiments at the time of writing, but we shall ignore this. It seems most likely that neutrinos make a small contribution towards Ω_M, but they cannot be the sought for cold dark matter.

3.12.2 Axions?

One might expect that the labelling of particles and antiparticles should be a matter of convention, so we could swap the labels consistently. This is not quite true. If only strong interactions were to exist then the statement could be made true by swopping handedness (viewing the Universe in a mirror) as well as swopping particle–antiparticle assignments. This is called charge–parity (or CP) symmetry. For the weak interactions even this is false and we are able to specify the difference between matter and antimatter in objective terms experimentally. (What we call the positron is the particle more likely to be produced in the decay of the long-lived neutral kaon particle.) So we find that CP symmetry is preserved in experiments involving the strong interaction, but not in those involving weak interactions. The axion is a hypothetical particle which has the role of preventing CP violation in quantum chromodynamics, the theory of the strong force. If axions exist they are expected to constitute a significant fraction of the cold dark matter. Their mass is expected to lie in the range 10^{-6}–10^{-3} eV/c^2, so their number density at our location in the Galaxy would be in the range 10^{12}–10^{14} cm^{-3}. These number densities are obtained from the estimated mass density of dark matter at our position in the galaxy which is $\sim 10^{-24}$ g cm^{-3} or 0.6 GeV/c^2 cm^{-3} (Gates *et al* 1995).

An inter-university collaboration in the United States has set up a search for axions. Their method of detection relies on the prediction that axions would be converted to photons in a strong magnetic field. They employ a resonant microwave cavity which can be tuned to frequencies which correspond to photon energies in the range 10^{-6}–10^{-5} eV. Thus they cannot explore the full mass range proposed for axions. If axions produced a narrow peak in the power spectrum of the radiation from the cavity will be detected. Preliminary results set limits but detect no axions (Hagmann *et al* 1998).

3.12.3   Neutralinos?

The neutralino is predicted by theories which go beyond current particle physics to incorporate supersymmetry (SUSY). In such theories each known particle has a superpartner. In most SUSY theories the neutralino is the lightest of the super-particles and is stable. The particle has a mass in the range 10 GeV/c^2 to 1 TeV/c^2 and it interacts weakly with ordinary matter. For illustrative purposes we will take the neutralino mass to be 100 GeV/c^2. The idea is that in the early Universe, when $kT \gg mc^2$, these particles were in equilibrium with the radiation through the reaction

$$N + \overline{N} \rightleftharpoons \gamma + \gamma \qquad \text{(or other particle–antiparticle pairs)}.$$

When the temperature T fell below mc^2/k the abundance of neutralinos fell exponentially with temperature until the annihilation rate fell below the expansion rate. In the absence of annihilation the comoving number density would be constant from that time onwards. For particles with a mass of the order of 100 GeV/c^2 and a weak force interaction strength their present density comes out to be $\Omega \sim 1$ (Jungman *et al* 1996), which is just what we are looking for. This result may be a coincidence. Nevertheless it is encouraging and together with the strong motivation for SUSY in particle physics makes the neutralino the favourite candidate for the postulated cold dark matter.

3.13   The search for WIMPS

We saw earlier that the mass density inferred for the dark matter halo at our location is about 0.6 GeV/c^2 cm^{-3}. Dividing this density by 100 GeV/c^2, say, the mass of the neutralino, gives a number density $n = 0.006$ cm^{-3}. As the Earth moves around the galactic centre it will sweep through the halo particles which will be moving in all directions like the atoms of a gas. So the flux nv of WIMPs at the Earth will be of the order of 10^5 cm^{-2} s^{-1}. As the WIMPS interact weakly with matter they will, for the most part, pass straight through the Earth, but occasionally one will scatter elastically off of an atomic nucleus causing it to recoil. This provides us with a direct method for detecting WIMPS. A WIMP moving with a relative velocity of about 200 km s^{-1} would impart a recoil energy to a germanium nucleus of about 20 keV (problem 16).

The event rate R per kilogram of germanium is given by the relation

$$R = \sigma F N$$

where σ is the interaction cross section, F the flux through the germanium and N is the number of germanium atoms in 1 kg. The cross section is not known exactly, but using $\sigma = 10^{-36}$ cm^2, which is at the high end of weak interaction cross sections, gives $R \sim 0.1$ day^{-1} kg^{-1}. This is probably an upper limit to the event rate that could be expected. These low event rates, and the small energy imparted to a recoiling nucleus, make the experimental detection of WIMPS by this method difficult. It is necessary to distinguish between the events sought for and background events produced by cosmic-rays and background radioactivity. A detector situated on the surface of the Earth would register more than 100 events kg^{-1} day^{-1} from cosmic-rays in the energy range of interest (Jungman *et al* 1996). So the need for shielding makes it necessary to go as deep underground as possible. Even then there is a gamma-ray background producing $\sim$2 events day^{-1} keV^{-1} that has to be distinguished from events induced by WIMPs. Fortunately the dark matter detection rate is expected to vary over the course of a year because of the changing velocity of the Earth about the galactic centre. This velocity will be in the range $v = 200 \pm 30 \cos 2\pi/3$ km s^{-1}, where 200 km s^{-1} is the speed of the Solar System in the Galaxy and 30 km s^{-1} is the speed of the Earth round the Sun in a plane making an angle of 60° to the galactic disc. Thus we look for events which have this characteristic variation.

The Cryogenic Dark Matter Search (CDMS) collaboration based in the United States uses a germanium detector. They cool very pure germanium crystals down to a temperature of 20 mK. At this temperature the heat capacity of the germanium is so low that an energy of a few keV is sufficient to raise the temperature by a measurable amount. Gamma-rays from radioactive decay will also heat the crystal, but they can be distinguished from nuclear recoils because they create more ionization than the recoils. Therefore by measuring both the heat energy and the ionization energy it is possible to distinguish between the two sorts of event. It is expected that a sensitivity of about 0.01 events day^{-1} kg^{-1} will be achieved within a few years.

A complementary Italian–Chinese collaboration, DAMA (from DArk MAtter), employs a large array of sodium-iodide detectors located 1400 m below ground in the Gran Sasso laboratory in Italy. The elastic recoil of a nucleus of sodium or iodine resulting from the impact of an incident particle produces scintillation photons that can be detected by photomultiplier tubes. DAMA has been detecting recoils that appear to satisfy the expected criteria for dark matter and after four years of collection of data the count rate appeared to show a seasonal modulation with the correct phase (a maximum in June). If the recoils detected were induced by dark matter then the corresponding WIMP mass lies between 44 and 62 GeV/c^2 and the event rate is 1 WIMP kg^{-1} day^{-1}. However, this result is in conflict with the CDMS collaboration which has not so far detected any events that can definitely be attributed to dark matter.

An indirect method of searching for WIMPS is to look for energetic neutrinos coming from the Sun (Press and Spergel 1985). A WIMP on its way through the Sun can become gravitationally bound to the Sun if it loses enough energy through scattering to bring its velocity below the escape velocity from the Sun. Once bound it can undergo further scatterings and sink to the Sun's centre. The concentration of the WIMPS accumulated in this way is determined by the equilibrium between the rate of capture and rate of annihilation. The WIMPS annihilate into quarks and leptons. Of the decay products only the neutrinos will escape from the Sun. The typical energy of these neutrinos will be about one-half of the WIMP rest mass, that is several GeV, so they are readily distinguishable from solar neutrinos which have energies of less than about 15 MeV. Some of the muon neutrinos that intercept the Earth will undergo a charged current interaction in the rock under a large neutrino detector. The resultant muons will register in the detector when it is on the opposite side of the Earth from the Sun, since the muons travel in the direction of the incoming neutrino. These muons provide a recognizable signature for WIMP annihilation neutrinos coming from the Sun. In a similar way WIMP annihilation could be occurring in the Earth. The current second generation neutrino detectors, such as SuperKamiokande, which are used for this work have an area of about $1000 \, \text{m}^2$. Detector areas much larger than this will be needed to make indirect detection as sensitive as the direct germanium detectors (Sadoulet 1999).

At the time of writing, although there is a strong belief in SUSY amongst theoretical particle physicists, there is no definite experimental evidence for it. The Large Hadron Collider (LHC), currently being built at CERN and due to start operating in 2005, will detect SUSY particles if they exist. If this happens then it would greatly strengthen the case that WIMPS are the non-baryonic dark matter.

3.14 Antimatter

Seen from a distance an antimatter star is indistinguishable from a star of matter, an antimatter Moon from one made of matter. From one point of view the Ranger spacecraft (or even the Apollo astronauts) may be regarded as (unnecessarily sophisticated) lunar antimatter detectors. The failure to observe spontaneous annihilation of these detectors can be taken as conclusive proof that the Moon is made of matter rather than antimatter. This mode of investigation is not at present capable of significant extension. Since we cannot go to the Universe, we must wait for it to come to us. As far as the present topic is concerned it does this in two ways: cosmic-rays and gamma-rays (Steigman 1976).

Cosmic-ray showers contain secondary antiparticles produced in collisions in the atmosphere which must be distinguished from primary sources. Antiprotons should arise from cosmic-ray collisions in the interstellar matter, but anti-helium nuclei must come from primary sources. Attempts to detect anti-helium 3 have yielded upper limits that put stringent constraints on the primary flux. For example, about 3×10^{-4} of the observed cosmic-ray flux could come from the Virgo Supercluster. If half of the Supercluster were made of antimatter, we would expect a proportion of 1.5×10^{-4} of antiparticles in the primary cosmic-ray flux. The observed fraction is much less than this allowing us to conclude that the Universe must be dominated by matter out to supercluster scales.

At the boundaries of matter and antimatter annihilations will produce gamma-rays. The upper limit to the gamma-ray background means that the

Universe cannot contain significant proportions of antimatter on less than supercluster scales and perhaps beyond.

The same effect on even larger scales will produce characteristic striation in the microwave background which the next generation of satellite experiments will be able to detect. Antimatter in the Universe will also be ruled out (or discovered) by cosmic-ray experiments in space.

3.15 Appendix. Derivation of the virial theorem

Here we derive the virial theorem for the case of N point particles acting gravitationally upon each other. Let the ith particle have mass m_i and position vector r_i relative to an arbitrarily chosen origin at rest. Then the equation of motion for the jth particle is

$$m_j \frac{d^2 r_j}{dt^2} = \sum_{i \neq j} \frac{Gm_i m_j}{|r_i - r_j|^3}(r_i - r_j), \tag{3.21}$$

where the right-hand side is the vector sum of the gravitational force on the jth particle due to all the other particles. Taking the scalar product with r_j, and summing over j, gives, on the left-hand side,

$$\sum_j m_j \frac{d^2 r_j}{dt^2} \cdot r_j = \sum_j m_j \frac{d}{dt}\left(\frac{dr_j}{dt} \cdot r_j\right) - \sum_j m_j \left(\frac{dr_j}{dt} \cdot \frac{dr_j}{dt}\right) \tag{3.22}$$

$$= \frac{d^2}{dt^2} \sum_j \frac{1}{2} m_j (r_j \cdot r_j) - \sum_j m_j \left(\frac{dr_j}{dt} \cdot \frac{dr_j}{dt}\right).$$

The first term on the right is the second derivative of the moment of inertia, which is zero if the system is in equilibrium (since no global properties are changing). The second term is just $-2T$. On the right-hand side of (3.21) we take the scalar product with $r_j = \frac{1}{2}(r_i + r_j) - \frac{1}{2}(r_i - r_j)$ to obtain

$$\sum_j \sum_{i \neq j} \frac{Gm_i m_j}{|r_i - r_j|^3}(r_i - r_j) \cdot (r_i + r_j) - \sum_j \sum_{i \neq j} \frac{Gm_i m_j}{|r_i - r_j|}. \tag{3.23}$$

The first term is zero by symmetry, since each term in the sum occurs twice with opposite sign: for example, $i = 1$, $j = 2$ and $j = 1$, $i = 2$ give cancelling terms. The second term in (3.23) is the gravitational potential energy of the system, V. Therefore the surviving terms in (3.22) and (3.23) yield the result $2T + V = 0$, as required.

3.16 Problems

Problem 13. *Compute the contribution of the background radiation to the density parameter at the present time, given the temperature of the radiation is 2.725 K.*

You may assume that the energy density of the radiation field is given by aT^4 where a is the radiation constant. Estimate the number density of background photons.

Problem 14. *The mass in the Coma cluster is approximately $10^{15} M_\odot$ within a radius of 2 Mpc. Estimate the radius of a comoving volume of matter at the mean density which contains the same mass as the cluster. What is the temperature of the gas in the cluster, assuming it is isothermal? What would be the corresponding density profile of the cluster mass? The x-ray flux from bremsstrahlung is 5×10^{37} W in the waveband 2–10 keV. What is the mass of hot gas in the cluster? Estimate roughly how long the gas would take to cool.*

Problem 15. *Show that equation (3.11) gives (3.13) for the Schecter luminosity function. (The Γ function is defined by $\Gamma(z) = \int_0^\infty t^{z-1} e^{-t} \, dt$.) What is the luminosity density ℓ for the parameters quoted in section 3.3.3?*

Problem 16. *Show that a neutralino moving with a relative velocity of about $200 \, km \, s^{-1}$ would impart a recoil energy to a germanium nucleus of about 20 keV.*

Chapter 4

Radiation

4.1 Sources of background radiation

The stars shine not out of darkness, but against a background of light from all the galaxies too faint to be seen as individual sources. Not just for visible light, but across the whole electromagnetic spectrum, we must distinguish between discrete emission from identified sources, and a diffuse or background radiation field. The background emission may come from faint discrete sources, in which case the distinction depends on the sensitivity of the detectors. Or there may be genuinely diffuse sources of radiation, for example from an intergalactic medium, for which the distinction would be real.

It is the diffuse radiation originating from well beyond our Galaxy that concerns us in this chapter. However, over much of the electromagnetic spectrum the diffuse radiation that we detect is dominated by local emission within the Galaxy. When this is the case we have to subtract off the radiation coming from the galactic foreground in order to obtain the cosmologically interesting background. This can be done successfully only when we have a good enough understanding of the local emission to model it reasonably accurately. Thus, in some regions of the electromagnetic spectrum we can as yet only estimate upper limits to the intensity of the cosmic background radiation. Figure 4.1 displays the background radiation intensity as a function of frequency across the electromagnetic spectrum. We outline here what is known about the different regions of this spectrum.

4.1.1 The radio background

In the radio region the major contribution to the background is provided by diffuse emission from the Galaxy. The Galaxy is permeated by a magnetic field of average strength 2×10^{-10} T. This is inferred from observations of the radio emission from pulsars since Faraday rotation (the frequency-dependent rotation of the plane of polarization), and dispersion (the frequency-dependent speed of

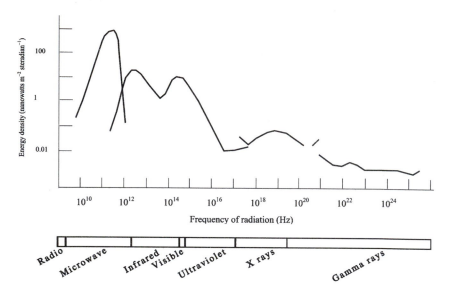

Figure 4.1. The energy density of the background radiation across the electromagnetic spectrum (from Hasinger 2000).

propagation) of the signals depend on the magnetic field strength along the path. Moving in this magnetic field are cosmic-ray electrons, with energies that go up to about 10^{12} eV per electron. These electrons, produced presumably by Galactic supernovae, interact with the magnetic field to emit synchrotron radiation at radio frequencies. The discovery of this radiation constituted the first extension of astronomy beyond the optical window. From independent data on the energy spectrum of the cosmic-ray electrons one can deduce a spectrum for this galactic radio emission, and the smaller extragalactic contribution to the radio background can then be determined by subtracting the galactic emission. When this is done a power law spectrum for the extragalactic intensity is obtained. It is given approximately by (Peacock 1999)

$$i_\nu = 6 \times 10^{-23} \left(\frac{\nu}{1\ \text{GHz}} \right)^{-0.8} \text{W m}^{-2}\ \text{sr}^{-1}\ \text{Hz}^{-1}.$$

This flux is most simply explained as the integrated emission of sources too weak or too distant to be seen individually. Known extragalactic discrete sources, such as radio galaxies and quasars, have appropriate spectra, and plausible extrapolation from the observed brighter sources yields the correct integrated intensity. The energy density of the radiation is about 10^{-20} J m^{-3} in the band 10^6–10^9 Hz and this is distributed throughout space. For comparison, note that the energy densities of the galactic magnetic field and of cosmic-rays in the Galaxy are both of the order 10^{-13} J m^{-3}.

In contrast, there are, as far as we know, no sources emitting significantly in the microwave region of the spectrum. Nevertheless, a flux of radiation with a blackbody spectrum at a temperature of about 2.7 K is observed. The absence of local sources and the isotropy of this radiation implies that it is coming to us from cosmological distances. As we shall see, the discovery of this radiation ranks in importance with the discovery of the expansion of the Universe. It stimulated the renaissance of physical cosmology in the 1960s and led to the emergence of cosmology from the vapours of speculation into a branch of physical science.

4.1.2 Infrared background

The cumulative emission from galaxies at all stages of their histories will give rise to an isotropic infrared background radiation ranging in wavelength from 10^{-2} cm (100 μm) to 10^{-4} cm (1 μm). In particular, the infrared background should contain highly redshifted light from a time when the first stars were forming. So the infrared background is of considerable interest to cosmology. However, detecting this background presents formidable challenges. Both the interplanetary space of the Solar System and the interstellar space of the Galaxy contain dust which absorbs starlight and re-radiates it into the infrared band. In order to measure the extragalactic infrared background it is therefore necessary to model this large local emission and then to subtract it from the measured total. This has now been done, using the measurements obtained with instruments on the Cosmic Background Explorer (COBE) satellite, and detections of the infrared background are reported at wavelengths of 140 and 240 μm with intensities $\nu i_\nu = 25 \pm 7$ nW m^{-2} sr^{-1} and 14 ± 3 nW m^{-2} sr^{-1} respectively (Hauser *et al* 1998, Puget *et al* 1996).

4.1.3 Optical background

At optical wavelengths the sky is again dominated by foreground emission from the Galaxy and only upper limits to the extragalactic background are available. The energy density, u, in the optical band of galactic starlight is 10^{-13} J m^{-3} giving a surface brightness, $i_\nu = u_\nu c/4 = 2 \times 10^{-21}$ W m^{-2} sr^{-1} Hz^{-1}. In contrast, the extragalactic surface brightness is estimated to be less than about 5×10^{-23} W m^{-2} sr^{-1} Hz^{-1} at a wavelength of 5000 Å. This, and corresponding upper limits at other wavelengths, restrict the possible numbers of faint galaxies. In particular, the limit at 2.2 μm is 7×10^{-23} W m^{-2} sr^{-1} Hz^{-1}, close to the integrated light of galaxies at this wavelength. This implies that a true diffuse component is small compared to the contribution from discrete sources. Surveys using CCD detectors reveal a population of faint blue galaxies of about 300 000 galaxies per square degree of sky with redshifts estimated to be up to about three (Tyson 1995). The UV emission from these galaxies, redshifted into the optical, gives an integrated spectral energy distribution that rises in the blue.

4.1.4 Other backgrounds

At the high-energy end of the electromagnetic spectrum the x-ray background has been measured between 0.1 keV (50 Å) and 1 MeV and the gamma-ray background beyond 1 MeV up to 10 GeV. In the last decade it has been established that the principal sources of this background are active galactic nuclei, i.e. quasars and Seyfert galaxies. We expand these comments in section 4.6.

The electromagnetic spectrum does not exhaust the possibilities for storing energy in the Universe in the form of radiation, by which we mean fields propagating at the speed of light or, equivalently, particles of zero rest mass. We must also consider zero mass neutrinos and gravitational waves, although neither has yet been detected in a cosmological context. In chapter 7 we shall show that the standard big-bang model predicts a background of relic neutrinos which have number density 336 cm^{-3} and an energy density equal to 0.68 times the energy density of the microwave background photons. At the present time the prospect of detecting these neutrinos directly is remote. In fact, it seems quite likely that neutrinos have a small rest mass. If this is less than $kT_\nu/c^2 \sim 2 \times 10^{-4}$ eV/c^2 for $T_\nu \sim 2$ K, which would be the temperature of a background of neutrino radiation at present, then the neutrinos can be treated as effectively massless. If they are more massive then they are non-relativistic at present and are treated as part of the matter density.

Gravitational waves are propagating distortions of the gravitational field. These are not possible in Newtonian theory, where gravity propagates instantaneously, but in relativistic gravity theories such disturbances propagate at the speed of light. Because gravity interacts so weakly the observational limits on the energy density of gravitational waves are actually quite weak. The main constraint comes from the distortions in the microwave background radiation that would be induced by intervening gravitational fields. These limit the contribution of the equivalent mass density in gravitational waves to the corresponding density parameter Ω_G to be less than 0.3 $(\lambda/1 \text{ Mpc})^{-2}$ on wavelengths greater than $\lambda = 1$ Mpc (Partridge 1995).

We shall consider the microwave and x-ray backgrounds in turn in the following sections. Since its discovery the cosmic microwave background has been by far the most important signal for cosmology. According to the simplest picture that we have for its origin, it is not a product of any astrophysical process currently operating. If we omit from our discussion the first 10^{-10} s of the history of the Universe, so that we confine ourselves to laboratory conditions and standard physical processes, then we must regard the microwave photons as part of the initial input into the Universe. This gives us the 'hot big-bang' model of cosmology which has proved so successful in accommodating our observations in a coherent picture. On this view the microwaves are signals from the 'creation' and much of our subsequent discussion will be concerned with the information they bring us. What happened before 10^{-10} s will be explored in chapter 8.

4.2 The microwave background

At wavelengths shorter than 50 cm the galactic emission is negligible and is thought not to rise again until radiation from dust becomes significant at a wavelength around 0.04 cm in the far infrared. Nor are the known discrete extragalactic sources expected to give any contribution in this microwave region. In the late 1940s Alpher and Herman, and Gamov, had predicted that thermal radiation with a temperature of 5–10 K, peaking at about 0.8–0.4 cm, would be the signal of nucleosynthesis in the hot big bang, but this had been largely forgotten. It therefore came as something of a surprise to Penzias and Wilson at the Bell Telephone Laboratories to find an apparently non-terrestrial source of radiation at a wavelength of 7.35 cm. They had started out by looking for the source of what they took to be excess noise in the Holmdel horn antenna, but were finally forced to conclude that the supposed noise was, in fact, a real signal coming from space (Penzias and Wilson 1965). This was less of a surprise some 30 miles down the road in Princeton, where Dicke had instigated the construction of a radiometer to look for just such a signal. Dicke had been investigating the irreversible production of photons in an oscillating Universe and it was his intention to look for these photons which, he conjectured, would be observed as a ubiquitous sea of microwave radiation at the present day. The correct interpretation of the microwave radiation was given by Peebles, who was working with Dicke, and appeared in a companion paper to that of Penzias and Wilson (Dicke *et al* 1965). For a fuller account of the history see Peebles (1993) and Weinberg (1977).

Atmospheric oxygen and water vapour emit microwaves so their emission must be subtracted from ground-based measurements in order to get the spectrum of the cosmic microwave background. At wavelengths less than about 1 cm these corrections are much larger than the signal so it is necessary to place detectors above the Earth's atmosphere to determine this part of the spectrum. The early measurements were carried out from the ground and were consistent with a blackbody distribution at a temperature of about 3 K. But they could not determine the position of the peak, which is at 0.2 cm, and the shape of the spectrum beyond. This situation changed dramatically in 1990 when the spectrum obtained from the COBE satellite was published (Mather *et al* 1990). Figure 4.2 shows the spectrum obtained with COBE's Far-Infrared Absolute Spectrometer (FIRAS). The FIRAS observers found, to the accuracy of the measurements, that their spectrum is a perfect fit to a blackbody at a temperature of 2.725 ± 0.001 K (Mather *et al* 1999).

4.2.1 Isotropy

That this radiation is non-local in origin is shown by its high degree of isotropy on large angular scales. Thus no association is found, for example, with the galactic plane, the Local Group or the Virgo supercluster. It follows that the source of the emission cannot be anything in our local environment. We can now argue that the radiation is universal, and further that it is homogeneously distributed. This

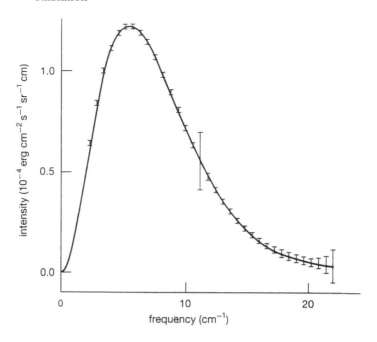

Figure 4.2. The spectrum of the cosmic microwave background radiation obtained by the COBE satellite (from *Physics World*, September 1996, p 31).

last step follows from the cosmological principle (chapter 1). This states that our location in the Universe is typical, and hence that observers at all other locations would also see a nearly isotropically distributed background radiation. Isotropy about each point then implies homogeneity.

Let us look at a piece of evidence in support of this assertion. We will see in section 4.4.2 that the expansion of the Universe does not change the blackbody nature of the spectrum but cools the radiation such that the temperature at the epoch corresponding to redshift z is given by

$$T(z) = T_0(1 + z), \tag{4.1}$$

where T_0 is the present temperature. In a Universe filled with a uniform sea of radiation this relationship applies everywhere. So if we measure the temperature of a cloud of gas, at redshift z, that is in thermal equilibrium with the background radiation, then its temperature should be given by equation (4.1). In fact this has been done for a cloud at a redshift of 1.776. The temperature found was 7.4 ± 0.8 K, in agreement with equation (4.1) which gives $T(z = 1.776) = 7.58$ K (Songaila *et al* 1994, see also Srianand *et al* 2000).

Having used the isotropy of the microwave background to establish its cosmological status, we can use the small but significant departures from isotropy to provide information about the departure of the Universe from uniformity. The isotropy of the radiation is investigated by comparing the temperatures of pairs of points on the sky. Such differential measurements are more accurate than absolute determinations of temperature. The first departure from isotropy was detected in the 1970s when it was found that the temperature varies cosinusoidally across the sky with the fractional departure from the mean temperature given by

$$\Delta T / T \approx 3.3 \times 10^{-3} \cos \theta. \tag{4.2}$$

Here θ is the angle between the direction of viewing and the hottest point on the sky. In section 4.5 we shall see that this variation arises from our motion with respect to the rest frame of the radiation and so is not intrinsic to the microwave background itself.

The dipolar variation in equation (4.2) refers to anisotropy over the whole sky. On intermediate scales of a few degrees, analysis of the first year's data from the Differential Microwave Radiometer (DMR) on board the COBE satellite revealed root mean square temperature fluctuations $\Delta T = 30 \pm 5 \ \mu K$ on angular scales of 7° (Smoot *et al* 1992). These fluctuations are believed to be intrinsic to the radiation and to be of fundamental importance. This subject will be treated in chapter 9.

4.3   The hot big bang

We show in the following section (section 4.4.2) that, once produced, a thermal radiation spectrum remains thermal as the Universe expands, although the temperature of the radiation decreases. Reversing this we see that the Universe was hotter in the past according to equation (4.1). In order to relate redshift to time we need a cosmological model (chapter 5), but we can say that if we extrapolate back to our currently assumed starting point, $t = 10^{-10}$ s, the Universe was hot and dense. Then the simplest explanation for the thermal background spectrum is that at 10^{-10} s the matter and radiation were in thermal equilibrium. This is the hot big-bang model. Is it true? In particular, can we extrapolate so far back or could the microwave photons have been produced more recently? Could we start from a cold big bang or even, as in the steady-state theory, no big bang at all? Even if the matter was hot at 10^{-10} s, were the matter and radiation in equilibrium or could they have been thermalized more recently?

The only natural way to produce a blackbody spectrum is to leave matter and radiation together undisturbed for long enough so that they can come into thermal equilibrium. It is clear that the contemporary Universe is not in thermal equilibrium, since not everything in our environment is even approximately at the 3 K of the background radiation. The reason for this is that the mean free path of a photon in the present Universe is so great that the matter and radiation are effectively independent of each other: if this were not the case we would be unable to see galaxies and quasars at high redshift. Thus, under the present conditions, the whole lifetime of the Universe is not 'long enough' to bring about thermal equilibrium.

If we go back to a temperature of about 3000 K, or a redshift of about 1100, the radiation is hot enough to ionize the neutral atoms. At this redshift the scale factor is of the order of one-thousandth of its present value (see equation (4.5)) and the galaxies would have overlapped. Therefore at this redshift the Universe contained no galaxies. It consisted instead of ionized gas (plasma) which interacted strongly with the radiation by Compton scattering of photons on free electrons. However, this interaction is not yet strong enough to bring about equilibrium between matter and radiation. For this we have to go back to at least $z \sim 10^4$. Even then, the equilibrium spectrum would not be blackbody. This is because a blackbody spectrum can only be established if photons can be freely created and destroyed by true emission and absorption processes. Compton scattering changes the frequency of a photon, but conserves photon number. The equilibrium establishes a balance between up-scattering and down-scattering of photons, but not a thermal spectrum. For that we have to go back even further to involve processes that create and destroy photons. There are two which can be important in different circumstances. One is bremsstrahlung (or free–free radiation), in which electrons are accelerated (or decelerated) by ions and as a result absorb (or emit) photons. The other is double Compton scattering, in which an electron scatters off a photon and emits (or absorbs) a second photon in the process. In either case these interactions are effective enough at around $z \sim 10^6$–10^7 to bring about thermal equilibrium. We can draw several conclusions.

(i)   As long as radiation was present in some form before $z \sim 10^7$ it will now have a thermal spectrum, provided there were no processes operating more recently that acted to distort the spectrum substantially.

(ii) Special senarios must be concocted to thermalize radiation more recently, for example in the steady-state theory (see later).

The first point, in principle, allows a cold big bang, with radiation generated before a redshift of 10^7, provided the processes that produce it do not distort the spectrum at lower redshifts. Dissipation of material motions by viscosity is an example of a process that could heat the Universe and produce radiation, but it falls foul of the proviso because the matter currents would imprint distortions on the spectrum. Other examples, where processes generate the radiation before 10^{-10} s, are best regarded as competitors to inflation (chapter 8).

The lack of detectable departures from a thermal spectrum puts a limit on any energy release between $z = 10^7$ and 10^3 of about 10^{-4} of the energy in the radiation (Wright *et al* 1994). We shall discuss potential processes in chapter 7. Finally, we are now in a position to see why the presence of a universal thermal radiation proved, as we stated earlier, to be fatal to the steady-state theory.

4.3.1   The cosmic radiation background in the steady-state theory

The steady-state theory is built on the perfect cosmological principle which asserts that the Universe looks the same at all times as well as from all positions. It achieves this by postulating the creation of new matter at a rate which just compensates for the effects of expansion; so the mean density is constant (see section 2.7). As the steady-state Universe is always the same, and has no beginning, it cannot appeal to an early hot phase to explain the blackbody spectrum of the cosmic background radiation. We have to find an ongoing process that generates the 2.7 K radiation. While our partial knowledge of the background spectrum could accommodate small departures from an exactly thermal spectrum, steady-state theorists were able to suggest that the cosmic brackground radiation is produced from starlight which has been absorbed by 'dust' grains and re-radiated. This reprocessing can produce an approximate blackbody spectrum over most of the waveband, but only by choosing the 'dust' to be cylindrical needles of graphite. How dust of this shape and composition is produced is not clear. This explanation is clearly very contrived. In any case the latest measurements of the cosmic microwave background spectrum by the COBE satellite do show that the spectrum is blackbody to a high accuracy and so appear definitively to rule out the steady-state model as well as several other non-standard models (Wright *et al* 1994).

4.4   Radiation and expansion

Physical cosmology deals with the interaction of matter and radiation in the arena of an expanding Universe. In this section we will present some results which describe the behaviour of blackbody radiation in such a system. First, we make the assumption that the interaction of the radiation with the matter can be neglected. As we have previously argued (section 4.3) this is a good approximation at the present time and back to the time corresponding to a redshift $z \sim 1000$. Paradoxically, it also turns out to be a valid description of the radiation for the early stages of the Universe, when matter and radiation were in thermal equilibrium. We will return to this point in chapter 7.

4.4.1   Redshift and expansion

Consider first the effect of expansion on the wavelength of light. Since photons do not carry markers which tell us where they come from, our conclusions will apply both to starlight and to the microwave background. Take a small region of the Universe containing background radiation but otherwise empty and surround it with a box with walls that are perfect mirrors inside and out. Let the box be free to expand with the Hubble flow. The presence of this box does not disturb the Universe, since the reflection simply swops a photon that would have been leaving the region of the interior of the box for one that would have been entering. The reflection does, however, introduce a Doppler shift

$$\frac{d\lambda}{\lambda} = \frac{dv}{c} = \frac{H\,dx}{c} = \frac{\dot{R}}{R}\,dt = \frac{dR}{R},$$

and hence

$$\lambda \propto R(t). \tag{4.3}$$

This equation says that the wavelength of a wave stretches with the Universe. In our box the expansion does not create additional nodes so a standing wave has nodes that move with the sides of the box. Note that while this result is strictly true this is not a genuine proof. It depends on the assumption that we can treat the box as expanding slowly in the sense that the timescale associated with the expansion, $1/H$, is large compared to the time it takes light to cross the box. Since the box must contain at least one wavelength this condition is $1/H \gg \lambda/c$. If the radiation in the box has temperature T then the peak of the spectrum is at $\lambda/c \sim h/kT$. Thus, for the validity of the argument we require $T \gg hH/k$, which is readily satisfied. Nevertheless, a full proof involves an analysis of Maxwell's equations for the radiation field, modified to take account of the expanding Universe (Ellis 1971).

From equation (4.3) we have

$$\frac{\lambda_0}{\lambda_e} = \frac{\nu_e}{\nu_0} = \frac{R(t_0)}{R(t_e)}. \tag{4.4}$$

Combining this result with equation (2.3), the definition of redshift gives

$$1 + z = \frac{R(t_0)}{R(t_e)}, \tag{4.5}$$

which is known as the Lemaître redshift relation. (See also section 5.8 where we derive this relation for relativistic cosmology.)

4.4.2 Evolution of the Planck spectrum

We can now show that a blackbody spectrum is maintained under expansion. We suppose that the spectrum is blackbody at time t. The Planck function tells us how many photons, $n_\nu \, d\nu$, per unit volume of space occupy a given frequency interval $d\nu$ at frequency ν in thermal equilibrium at temperature T, so

$$n_\nu \, d\nu = \frac{8\pi}{c^3} \frac{\nu^2 \, d\nu}{\exp(h\nu/kT) - 1}. \tag{4.6}$$

As such a radiation field expands, the change in frequency relabels the modes. Note also that, under the assumption that we are neglecting the interaction of radiation and matter, no new photons are created and none are destroyed. Therefore, at a later time t_0, the number of photons in the mode labelled by $\nu(t_0)$ in a given comoving volume of space equals the number originally in the mode labelled by $\nu(t)$, where, from equation (4.4)

$$\nu(t_0) = \nu(t) \frac{R(t)}{R(t_0)}. \tag{4.7}$$

Since the given region of space increases in volume by a factor R_0^3/R^3, where, for simplicity, we have written $R = R(t)$ and $R_0 = R(t_0)$, we have for the number density of the photons at time t_0,

$$n_{\nu_0}(t_0)\, d\nu_0 = \frac{R^3}{R_0^3} n_\nu\, d\nu.$$

Now on substituting for $n_\nu\, d\nu$ from equation (4.6) we obtain

$$n_{\nu_0}(t_0)\, d\nu_0 = \frac{R^3}{R_0^3} \frac{8\pi}{c^3} \frac{\nu^2\, d\nu}{\exp(h\nu/kT) - 1}.$$

Finally by using equation (4.7) this can be written

$$n_{\nu_0}(t_0)\, d\nu_0 = \frac{R^3}{R_0^3} \frac{8\pi}{c^3} \left\{ \frac{R_0^3}{R^3} \frac{\nu_0^2\, d\nu_0}{\exp(h\nu_0 R_0/RkT) - 1} \right\}$$

$$= \frac{8\pi}{c^3} \frac{\nu_0^2\, d\nu_0}{\exp(h\nu_0 R_0/RkT) - 1}.$$

This is a Planck spectrum with temperature $T_0 = RT/R_0$. So expansion maintains the blackbody form of the spectrum but lowers the temperature in such a way that

$$RT = \text{constant}. \tag{4.8}$$

Note that since expansion maintains thermal equilibrium it is a reversible process from the point of view of thermodynamics.

4.4.3 Evolution of energy density

As a simple application of the results of the last section consider how the energy density of the radiation changes with time. Since we are dealing with blackbody radiation, the energy density is $u = aT^4$ and so, from equation (4.8),

$$uR^4 = \text{constant}. \tag{4.9}$$

Note that this result is consistent with the maintenance of a Planck spectrum and the relabelling of modes, as indeed it must be, for we have

$$u_\nu\, d\nu \propto \frac{\nu^3\, d\nu}{\exp(h\nu/kT) - 1} \propto \frac{1}{R^4} \frac{\nu_0^3\, d\nu_0}{\exp(h\nu_0/kT_0) - 1}.$$

Therefore the energy density in each band $d\nu$ is proportional to R^{-4}, and so is the total over all bands.

Since the volume, V, of a given region of space is proportional to R^3, this means that the total energy of radiation in this region decreases as R increases. The change in internal energy is

$$d(uV) = d(u_0 V_0 R_0/R) = -(u_0 V_0 R_0) \frac{dR}{R^2} = -uR^2(V_0/R_0^3)\, dR, \tag{4.10}$$

where $V = (R/R_0)^3 V_0$ is the volume at time t in terms of that at t_0 and we have used equation (4.9). The radiation exerts a pressure which does work as the Universe expands. The work in expanding by dR is

$$p_r \, dV = \tfrac{1}{3}u \, d(R^3)(V_0/R_0^3) = R^2 u \, dR(V_0/R_0^3),$$

where $p_r = (1/3)u$ is the radiation pressure. Thus the work done is equal to the change in internal energy. By application of the second law of thermodynamics to the radiation this means that the heat input into the radiation must be zero so the expansion is adiabatic. We now confirm this by calculating the entropy of the radiation which should be constant.

4.4.4 Entropy of radiation

Let S be the entropy and U the energy of a system in a volume V. Then the second law of thermodynamics implies

$$dU = T \, dS - p \, dV$$

so

$$dS = \frac{dU}{T} + \frac{p \, dV}{T} = \frac{V}{T}\frac{du}{dT} \, dT + \frac{(u + p)}{T} \, dV$$

where $u = U/V$ is the energy density. It follows that

$$\left(\frac{\partial S}{\partial V}\right)_T = \frac{u + p}{T}.$$

Since u, p and T are independent of V, the right-hand side is a constant. Integrating we get

$$S = \frac{(u + p)}{T}V + f(T).$$

Since S is an extensive variable (proportional to the volume of a substance), we must have $f(T) = 0$. Applying this to the radiation field, using $p = p_r = (1/3)u = (1/3)aT^4$, this gives an entropy density

$$s = (4/3)aT^3$$

for the entropy density of the radiation. The total entropy in volume V is $sV \propto T^3 V$, and from equation (4.8), this is constant as the Universe expands. This is the required result.

   Note that both the entropy and the number of particles in a comoving volume are constant, or, equivalently, the entropy density s and the number density of baryons n_b are both proportional to R^{-3}. Therefore the dimensionless ratio s/kn_b, the entropy per baryon, is a constant in time, i.e. it is a parameter that characterizes the Universe.

A Universe filled with radiation expands adiabatically as a gas of photons. Since $p \propto u R^4$ and $V \propto R^3$, we have $pV^{4/3} = $ constant, so the gas has adiabatic index 4/3, which is just the usual result for a photon gas.

The thermodynamics of the expanding universal radiation field is therefore like that of radiation imprisoned in a box with perfectly reflecting walls that expands adiabatically. In both cases the entropy is conserved and the internal energy decreases. There is one difference however. In the case of the box the internal energy has been lost through doing work on its surroundings and so can be accounted for. In the case of the Universe each comoving volume loses energy by doing work, but we cannot say that this work is being done on its surroundings because those surroundings are also losing energy in an exactly similar way: so we cannot say that the energy lost by one volume reappears in a neighbouring volume. Nor can we say that the radiation pressure is accelerating the expansion, hence converting internal energy to bulk energy because the uniform pressure means that there are no pressure gradients, hence no net forces. The resolution appears to be that we do not have a law of global energy conservation in cosmology. See, for example, Harrison (1995) and Peebles (1993).

Note that no assumptions about the laws of gravity have been made in this section, so the conclusions are valid generally for an expanding Universe.

4.5   Nevertheless it moves

Once we have convinced ourselves that the microwave background has cosmological significance, its properties can provide us with information about the Universe, a theme to which we shall find several occasions to return. Here we discuss how the detection of a dipole anisotropy in the background (Smoot *et al* 1977) has been used to determine the motion of the Earth (thereby confirming Galileo's comment on his famous recantation, which provides the title of this section).

To understand the origin of this dipole anisotropy we have to study how a thermal radiation field, which is isotropic to one observer, appears to a second observer moving with uniform relative velocity v. This is a problem in special relativity. We will choose the frame in which the radiation is isotropic to be the S frame and the frame of the moving observer to be the S$'$ frame , where the reference frames are in the usual configuration with x-axes aligned and the motion of S$'$ along the x-axis of S. Let the telescope of the observer point at an angle θ' with respect to his/her direction of motion. Now apply the energy momentum transformation

$$E = \gamma(E' + vP')$$

to the photons entering the telescope. This gives

$$\nu = \nu'\gamma(1 - \beta \cos\theta'), \tag{4.11}$$

where we have used $E = h\nu$, $E' = h\nu'$, $P' = -(h\nu'/c)\cos\theta'$ and $\beta = v/c$. The specific intensities i_ν and $i'_{\nu'}$ in the frames S and S' respectively obey the relationship

$$\frac{i'_{\nu'}}{(\nu')^3} = \frac{i_\nu}{\nu^3} \tag{4.12}$$

(see Rybicki and Lightman 1979, ch 4). As the radiation has an isotropic blackbody spectrum in S,

$$i_\nu = \frac{2h}{c^3}\frac{\nu^3}{\exp(h\nu/kT) - 1}. \tag{4.13}$$

We obtain, on substituting (4.13) into (4.12) and using (4.11),

$$i'_{\nu'} = \frac{2h}{c^3}\frac{\nu'^3}{\exp(h\nu'\gamma(1 - \beta\cos\theta')/kT) - 1},$$

which is just a blackbody distribution with temperature

$$T' = \frac{T}{\gamma(1 - \beta\cos\theta')}.$$

Thus the radiation seen in S' has a blackbody spectrum but with a temperature which depends on the angle between the line of sight and the x'-axis. This is not surprising as a system in thermal equilibrium must appear so to all inertial observers.

For $\beta \ll 1$ this becomes, on using the binomial expansion to first order,

$$T' = T(1 + \beta\cos\theta').$$

Thus the microwave sky should appear hottest in the direction of motion, $\theta' = 0$, and coolest in the opposite direction, $\theta' = 180°$ with a dipole variation of the form

$$\frac{\Delta T}{T} = \frac{v}{c}\cos\theta', \tag{4.14}$$

where $\Delta T = T' - T$. Equation (4.14) has the same form as equation (4.2) so we interpret the observed dipole as arising from our motion with respect to the rest frame of the radiation.

Observations of $\Delta T/T$ can therefore be used to find our velocity, v, with respect to the rest frame of the microwave radiation, which acts as a universal rest frame.

4.5.1 Measurements of motion

Measurements made from the COBE satellite adjusted to the frame of the Sun give

$$T = (3.372 \pm 0.014) \times 10^{-3}\cos\theta' \text{ K} \tag{4.15}$$

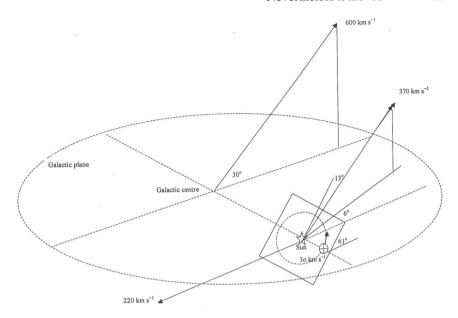

Figure 4.3. The motion of the Earth (adapted from Muller 1978). The velocity measured relative to the microwave background is shown by the double arrow.

(Fixsen *et al* 1996). Equations (4.14) and (4.15) yield 371 ± 1 km s^{-1} for the peculiar velocity of the Solar System with respect to the cosmic microwave background in the direction $l = 264.14° \pm 0.3$, $b = 48.26° \pm 0.3$. This velocity is the vector sum of the velocity of the Solar System about the centre of the Galaxy, which is some 220 km s^{-1}, the velocity of the Galaxy with respect to the centre of mass of the Local Group, roughly about 80 km s^{-1} and the velocity of the Local Group with respect to the rest frame of the microwave background. It might be anticipated that this last velocity would turn out to be small so it came as a surprise in the 1970s when it was discovered that the Local Group is moving with a velocity of about 600 km s^{-1}. Figure 4.3 shows how this large velocity for the Galaxy is obtained.

This velocity is thought to arise mainly from the gravitational attraction of a large concentration of mass, dubbed the Great Attractor, which lies in the part of the sky behind the dense central region of the Milky Way—the zone of avoidance. A recent study of this region by Kraan-Korteweg *et al* (1996) has identified a rich cluster with a mass comparable to that of Coma and located at a distance of about 80 Mpc from us. This cluster A3627 together with other clusters in the same part of the sky might fully account for our large velocity.

The frame of reference in which the microwave background appears isotropic can be regarded as providing a universal standard of rest. This sometimes leads to concern that there might be a conflict with the special theory of relativity, since this asserts the impossibility of establishing an absolute rest frame. What relativity actually forbids is the determination of motion by local experiments, i.e. experiments carried out in a closed laboratory without reference to anything outside, like the Michelson–Morley experiment which looked for changes in the speed of light to reveal our motion with respect to a postulated absolute state of rest. There is no suggestion that the laws of physics single out the microwave background frame. The anisotropy observations measure our velocity relative to the microwave background and this, in principle, is no different from observing our motion relative to the stars.

4.6   The x-ray background

The spectrum of the x-ray background extends from an energy of 0.1 keV to about 1 MeV. This range is subdivided into the soft x-ray band (0.1–2 keV) and the hard x-ray band. The gamma-ray background covers energies beyond 1 MeV.

Figure 4.1 shows the spectrum of the x-ray and gamma-ray background. The spectrum appears smooth and non-thermal in character which suggests that the x-ray background cannot have originated with the microwave background in the thermal equilibrium conditions of the early Universe. It must therefore have been emitted more recently from astrophysical sources. The fact that we detect discrete x-ray sources, e.g. quasars, at high redshift tells us that the x-ray background is coming from sources at all epochs since the first luminous structures started to form at the end of the so-called dark age. A reason for supposing that much of the background is coming from large redshifts is that the x-ray sky is isotropic to a few percent on large angular scales once the local galactic emissions and identifiable discrete sources are removed. The only detected large-scale anisotropy is a dipole pointing in the approximate direction $l = 280°$, $b = 30°$ which, within the experimental errors, coincides with the direction of the microwave background dipole. This suggests that the dipole originates in our motion with respect to the distant sources. Under this assumption our velocity turns out to be 475 ± 175 km s^{-1} (Miyaji and Boldt 1990), which is roughly consistent with the value of 371 km s^{-1} obtained from the microwave background dipole. (See section 4.5.1.)

The energy range below about 1 keV is dominated by emission from a local hot plasma originating from supernova remnants and possibly from hot gas within the local group of galaxies; this latter is the intracluster gas referred to in section 3.8. Above an energy of 1 keV the cosmic extragalactic background dominates over x-ray emission from local sources.

The spectrum in the range 3–60 keV can be fitted with the following expression

$$Ei_E = 7.877 E^{0.71} \exp\left(-\frac{E}{41.13 \text{ keV}}\right) \text{ keV s}^{-1} \text{ sr}^{-1} \text{ cm}^{-2},$$

where E is the photon energy in keV. This spectrum is approximately thermal in shape as it rises to a peak at about 30 keV and then falls with increasing energy. This originally led to the suggestion that this part of the spectrum was produced by a hot diffuse intergalactic medium. The idea is that at some point after the formation of atoms at $z \approx 1100$ the neutral gas was reionized and heated to a high temperature by photons emitted in the formation of the first stars; this would have been at $z \sim 5$. Such a hot plasma would be a strong source of x-rays. However, another effect of the plasma would be to Compton scatter microwave background photons and so distort their spectrum from a thermal distribution. Measurements of the microwave background spectrum with the COBE satellite found no evidence for departures from a blackbody distribution and show that, at most, a uniform hot intergalactic medium could produce only about 10^{-4} of the x-ray background flux (Wright *et al* 1994). Thus the diffuse source idea is now ruled out.

Let us now examine the alternative hypothesis that the x-ray background comes from discrete sources. At the soft x-ray end of the spectrum, i.e. below 2 keV, the extragalactic background has been nearly completely resolved into individual sources. Most of these sources are active galactic nuclei (AGN), which are point sources at the centres of some galaxies, thought to be powered by accretion on to black holes. However, there is also a significant contribution from galaxy clusters with their hot intracluster gas and also from narrow emission line galaxies (Treyer *et al* 1998). At energies above a few keV it seems likely that the background comes from unresolved AGN alone. In order to test this idea we need to know the x-ray spectra of the various classes of AGN, their luminosity functions, which tell us the spatial density of these classes as a function of luminosity, and finally we need to know how the AGN evolve in time. With this information it would be possible to model the x-ray background by summing up the fluxes from all the AGN over cosmic time. At present our knowledge of these quantities is incomplete. Nevertheless several different models have been constructed which provide a good fit to the measured spectrum from about 5 keV to about 100 keV (Comestri *et al* 1995). So the integrated emission of AGN can account for the spectrum and flux of the x-ray background over, at least, a part of its range. However, as several different models employing different assumptions give equally good fits to the spectrum we cannot claim that the problem of the production of the x-ray background is solved yet.

To account for the observed flux of the x-ray background its sources must extend back to a redshift $z \sim 5$, so much of the radiation is coming to us from high redshifts. Thus the x-ray background radiation enables us to probe fluctuations in the matter density at scales intermediate between those probed by the microwave background radiation, several thousand Mpc, and the scales probed by large galactic surveys, several hundred Mpc. The isotropy of the background radiation, therefore, tells us about the fluctuations in the matter density on scales $600h^{-1}$ Mpc: the root-mean-square fluctuations in the matter density on these scales are less than 0.2% (Treyer *et al* 1998). The small amplitude of $\delta\rho/\rho$ on these scales is strong evidence in favour of the cosmological principle (Wu *et al* 1999).

4.7 Problems

Problem 17. *Show that starlight gives an energy density of optical radiation in the Galaxy of order 10^{-13} J m^{-3} and a corresponding surface brightness of about 2×10^{-21} W m^{-2} sr^{-1} Hz^{-1}. Show that the equivalent mass density of the microwave background radiation is 4.6×10^{-31} kg m^{-3}.*

Problem 18. *Show that the spectrum for a distribution of massless fermions*

$$n(E)\,dE = \frac{4\pi g}{c^3}\,\frac{E^2\,dE}{\exp(E/kT)+1}$$

(g = constant) is preserved by expansion.

Problem 19. *Estimate the entropy per baryon in the Universe. What is the total entropy of the observable Universe?*

Problem 20. *The spectral surface brightness i_ν of radiation from a source with volume emissivity $j_\nu(t)$ at time t is given by*

$$i_\nu = \int dt \, c j_{\nu(1+z)}(t)(1+z)^{-3},$$

where, for thermal bremsstrahlung at temperature T,

$$j_\nu = 5 \times 10^{-40} n_e^2 T^{1/2} \exp\left(\frac{h\nu}{kT}\right) \text{ W m}^{-3} \text{ sr}^{-1} \text{ Hz}^{-1}.$$

To estimate the integral assume that all the emission comes from a single redshift (so j_ν is multiplied by a delta-function $\delta(z - z_0)$) and that the relation between redshift and time can be found using $R(t) = R(t_0)(t/t_0)^{2/3}$ which is appropriate to the Einstein–de Sitter model with a critical density, $\Omega = 1$ (chapter 5). Hence show that bremsstrahlung emission from a hot intergalactic medium at 10^7 K with density parameter Ω_{igm} is

$$i_\nu = 4 \times 10^{-26}(H_0/100 \text{ km s}^{-1} \text{ Mpc}^{-1})^5 \Omega_{igm}(1+z_0) \text{ W m}^{-2} \text{ s}^{-1} \text{ sr}^{-1} \text{ Hz}^{-1},$$

at frequencies $\nu \ll kT/h$. By comparison with the known x-ray sky brightness at 3 keV show that a diffuse intergalactic plasma could account for the observed x-ray emission. (However, the theory is incompatible with the pressure in Lyα forest clouds; the diffuse x-ray background is almost certainly attributable to active galaxies.)

Problem 21. *The reaction*

$$p + \gamma \rightarrow p + \pi^0$$

occurs between sufficiently energetic cosmic-ray protons and microwave photons. By considering a head-on collision, show that, for the reaction to occur, the Lorentz γ factor of the incident proton, mass m_p, must exceed

$$\gamma \simeq \frac{mc^2}{2Q},$$

where $m = 135\ MeV/c^2$ is the mass of the pion, Q the energy of the microwave photon and we have used the approximation $2m_p/m \gg 1$. Estimate the threshhold energy.

Chapter 5

Relativity

5.1 Introduction

As with many ideas that do not fit the prevailing views, the expansion of the Universe was discovered several times before it was generally recognized. In 1917 Einstein had applied his newly completed general theory of relativity to the distribution of matter in the Universe. In order to obtain a static distribution of matter, which seemed at the time to be a self-evident necessity, he added the so-called cosmological term to his original field equations. This term gives rise to a repulsion which can be adjusted to balance the gravity of the matter distribution. In the same year de Sitter published another solution to Einstein's equations for a vanishingly small density of matter. This was shown to predict a relation between redshift and distance, but was nevertheless interpreted as a second static solution. Throughout the 1920s it was taken for granted by cosmologists that the Universe was static, and that the main issue in cosmology was to decide between the solutions of Einstein and de Sitter.

On the observational side, by 1925 Slipher had obtained the spectra of 45 galaxies and had found that they showed a preponderance of redshifts. Using the 100 inch Mount Wilson telescope, Hubble obtained further galactic redshifts. He obtained the distances to these and to Slipher's galaxies and in 1929 announced 'a roughly linear relationship' between redshift and distance.

Although Hubble did not propose the idea of an expanding Universe in his 1929 paper it provided the first convincing evidence for an increase of galactic redshift with distance. In the following year, at the January meeting of the Royal Astronomical Society in London, Eddington and de Sitter reached the conclusion that neither of the static solutions was satisfactory and proposed looking for non-static solutions, giving, as Eddington described it later, a little motion to Einstein's inert matter or a little matter to de Sitter's prime mover. In fact, such solutions had already been found by Lemaître, who immediately communicated the fact to Eddington. From this point the paradigm shift from the static to the expanding Universe was rapid. Later it was discovered that in 1922 Friedmann had also found expanding solutions of Einstein's equations, and had communicated the fact to Einstein who at first incorrectly dismissed them as erroneous and then as physically uninteresting. Finally, Robertson and Walker independently proved that there are no further alternative solutions in which matter is distributed uniformly. As a result, the expanding Universe solutions of Einstein's theory are called the Friedmann, Lemaître, Robertson–Walker, models (or by their initials, FLRW or by various subsets of these originators or initials).

This chapter will explain the nature of these models, starting with what it takes to be a model in this context. In the next chapter we shall confront the models with observation.

5.2 Space geometry

The geometrical properties of space are Euclidean if (amongst other things) the Pythagorean law holds for the distance δs between nearby points (x, y) and $(x + \delta x, y + \delta y)$, namely

$$\delta s^2 = \delta x^2 + \delta y^2. \tag{5.1}$$

This is geometry as we intuitively understand it. Usually we consider the limit as the separation of the two points goes to zero and write

$$ds^2 = dx^2 + dy^2$$

instead of (5.1).

The question of the existence of non-Euclidean geometries amounts to the question of whether consistent alternatives to (5.1) are possible. An obvious suggestion is the distance law on a sphere of radius r,

$$ds^2 = r^2 \, d\theta^2 + r^2 \sin^2 \theta \, d\phi^2, \tag{5.2}$$

for which the angles of a triangle exceed 180°. It is important to realize that the intuitive picture of the surface of a sphere is misleading here. This picture forces a visualization of a sphere as an object embedded in an enveloping three-dimensional space. The question of the consistency of alternative geometries addresses the issue of the intrinsic geometry of a surface, independently of the properties it inherits from the surrounding space, and that is a non-trivial problem. In this context one can also address the question of alternative three-dimensional geometries, despite the fact that we cannot form a mental picture of these in a higher-dimensional environment.

Note that equation (5.2) is quite different from the expression in polar coordinates for a distance in Euclidean space. Equation (5.2) cannot be transformed into (5.1) by a change of coordinates. To put this another way, the value of the distance δs^2 between two fixed points in (5.1) is independent of the choice of coordinates used to obtain it (i.e. it is an invariant). The *form* of (5.1) is unchanged under a rather more restricted set of transformations of coordinates, namely rotations and translations.

On the other hand (problem 22), we *can* transform (5.1) into the form

$$ds^2 = dr^2 + r^2\, d\theta^2.$$

This does describe the same geometry as (5.1), but in different coordinates.

5.3 Relativistic geometry

Special relativity is all about time. In the special theory we define a proper time $\delta\tau$ between events at (t, x) and $(t + \delta t, x + \delta x)$ as

$$c^2\delta\tau^2 = c^2\delta t^2 - \delta x^2. \tag{5.3}$$

By analogy with the Pythagorean law we say that this defines a geometry of spacetime. As before we usually take the limit as the separation of the points goes to zero and write

$$c^2\, d\tau^2 = c^2\, dt^2 - dx^2. \tag{5.4}$$

The basis of kinematics in special relativity is that the proper time is a measurable quantity (independent of the motion of the observer, i.e. a Lorentz invariant). To this we add the observation that light-rays move on paths of zero proper time (i.e. $\delta\tau = 0$ for a light-ray).

For dynamics we need also some equations of motion. For a structureless particle of mass m subject to a force F and moving with speed v the motion is determined by

$$m\frac{d^2x}{d\tau^2} = \gamma F,$$

where $\gamma = (1 - v^2/c^2)^{-1/2}$. This equation is integrated to give $x(\tau)$. To get back to a description of the path of the particle in terms of $x(t)$, which is how we usually think of it, we use (5.4) to relate τ and t.

General relativity is all about two things: first proper time, then gravity. The most remarkable fact is that these turn out to be different aspects of the same thing. The idea that connects them is called the principle of equivalence.

5.3.1 The principle of equivalence

To an accuracy of at least one part in 10^{13} all bodies subject to no non-gravitational forces are found to fall at the same rate in a gravitational field. The key assumption that this holds exactly is referred to as the *universality of free fall*.

Consider now two identical laboratories, one floating freely in space remote from any gravitational influences and the other in free fall towards a massive body, the Earth say. By free fall we mean that the laboratory is subject to no non-gravitational forces. According to what is now known as the weak principle of equivalence, it is impossible by means of mechanical experiments for the occupants of the laboratories to distinguish between their situations. This follows because, for the occupants of both laboratories, bodies remain at rest, or in uniform motion, unless acted on by a force. In the first case (remote laboratory) this is because the laboratory is a Newtonian inertial frame. In the second case (free plunge to the Earth) it is a consequence of the universality of free fall, since bodies falling with the *same* acceleration remain relatively at rest (or in uniform relative motion). There is one proviso: that the laboratories are not extensive enough to detect the convergence of radially falling particles towards the centre of the Earth or the time of observation long enough. These statements are therefore true *locally* (i.e. in infinitesimal regions of spacetime).

If we remove this restriction to mechanical experiments, we obtain the strong principle of equivalence, according to which, no non-gravitational experiments at all can distinguish the two situations. This is less well tested. For example, it is know to be true for electromagnetic experiments to about 1 part in 10^3. But under certain general assumptions it can be deduced from the weak version of the principle. Finally, Einstein's principle of equivalence declares that no experiment whatsoever, even one involving gravity, can be used to distinguish the two cases (Will 1993).

Why is this so important? Because it is the physical basis of the general theory of relativity. The principles of equivalence mean that the effects of gravity are removed locally for freely falling observers. This means that, in principle, the effects of gravity can be deduced by transforming the many local observations back to a single global view. (This is where the complexities of the mathematics of relativity come in.)

5.3.2 Physical relativity

We start from the proper time locally of a freely falling observer. This must be given by (5.3) or (5.4) because locally free fall has removed the effects of gravity. In a global (non-inertial) frame, we expect to have a more general expression for the proper time. Thus, the basic idea is that in general relativity spacetime is allowed to have a more general geometry. Continuing with our simplification to one space dimension, this means that the proper time is given by an expression of the form

$$c^2 \delta \tau^2 = f(x, t) c^2 \delta t^2 + 2h(x, t) \delta x \delta t + g(x, t) \delta x^2 \qquad (5.5)$$

for some functions f, g and h. We have continued to use x and t for the coordinates, even though these are not now the same as the local (x, t) of (5.3), to avoid a proliferation of symbols.

The principle of equivalence tells us how the paths of light-rays and particles would be seen by freely falling observers in any gravitational field. We therefore expect these paths to be governed by (5.5) in a general frame of reference, since they are determined by (5.3) for the freely falling observers. It turns out that we require the proper time to be stationary with respect to variations in the path;

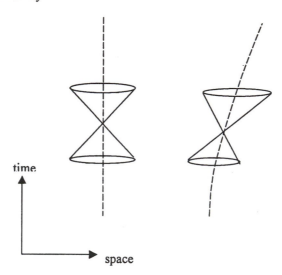

Figure 5.1. The tilting of light cones in curved spacetime. We can choose a frame of reference such that the geometry appears flat along any one trajectory, but, if gravity is present, the light cones on any other spacetime path will appear distorted.

the details need not concern us here. For light-rays we require in addition that $\delta\tau^2 = 0$. Hence the functions f, g and h *determine* the motion of light-rays and of particles in the presence of (other) matter, i.e. they represent the effects of gravity.

The expression (5.5) is often called the *metric* of spacetime (although this term strictly refers to the set of coefficients f, g and h) or the *line element*.

In *general* we cannot transform (5.5) into the form (5.4) by a change of coordinates, except in a single, arbitrary, infinitesimal patch. Figure 5.1 illustrates this. In general, therefore, the two metrics represent different geometries (i.e. different proper times), not just the same geometry in different coordinates. In such cases we say the spacetime of (5.5) is curved. Otherwise the spacetime is said to be flat. (So in special relativity, i.e. in the absence of gravity, spacetime is flat.)

For a general distribution of matter there are some complications. First the real world has two more dimensions. The expression for the metric or line element then involves a 4×4 symmetric matrix (g_{ij}). This makes it difficult to write out the equations of relativity term by term, so one has to invest in learning the relevant mathematical techniques (tensor calculus). Second, to derive, or even to state, the Einstein field equations for the way in which the functions f, g and h (or really the 10 components of g_{ij}) are determined by the matter distribution is complicated. However, if we are interested only in a homogeneous isotropic

Universe, and if we are willing to forgo the derivations, the results can be simply stated, and understood, with just these preliminaries, as we shall see.

5.4   Isotropic and homogeneous geometry

Our first task is to find the most general spacetime metric for a Universe which is spatially homogeneous and isotropic, i.e. one that satisfies the cosmological principle at all times. Our point of departure is the expanding currant loaf analogy introduced in section 2.2. This prompts us to propose the following metric:

$$ds^2 = c^2 \, dt^2 - R(t)^2 (dx^2 + dy^2 + dz^2), \qquad (5.6)$$

where, in terms of our previous notation, $ds = c \, d\tau$, and where $R(t)$ is some increasing function of time t.

Consider a pair of points having fixed space coordinates separated by δx, δy, δz. The spatial distance between these points at time t is

$$R(t)(\delta x^2 + \delta y^2 + \delta z^2)^{1/2}. \qquad (5.7)$$

So, apart from an overall constant, spatial distances at any fixed time are given by the Pythagorean metric. Therefore the metric (5.6) describes a geometry in which the three-dimensional space (x, y, z) at any time t is Euclidean. The spaces of constant time are therefore homogeneous and isotropic.

The factor $R(t)$, in (5.7) ensures that the spatial distance between the points increases with time. Therefore the spacetime geometry described by (5.6) is that of an expanding space. The fact that $R(t)$ is a function of time only, and not of position, ensures that the expansion is uniform and in accord with the cosmological principle. In fact $R(t)$ is just the scale factor, which we introduced in chapter 2, because it describes how the distances in space scale with time.

Observers having fixed position coordinates in this spacetime are, by symmetry, not acted on by any non-gravitational forces. Yet the distance between them increases with time. Since force-free particles do not recede from each other in the flat spacetime of special relativity, the four-dimensional geometry described by this metric is not flat, so the spacetime geometry must be curved. If R is a constant then the metric (5.6) is just the spacetime metric of special relativity: the constant factor R can be absorbed into the x, y, z by a transformation of coordinates and the distance scale redefined.

The question now arises as to whether the metric of equation (5.6) is the most general geometry that satisfies the cosmological principle. This is equivalent to asking whether *space* can be anything other than Euclidean and still satisfy the requirements of homogeneity and isotropy. We know that in two dimensions the Euclidean plane is a homogeneous and isotropic space. So too is the surface of a sphere, since it has no preferred points or directions. A 'sphere' of imaginary radius, which obviously cannot be visualized as a surface embedded in three-dimensional Euclidean space, is another example. In fact, these three possibilities complete the list of two-dimensional isotropic geometries.

For three-dimensional spaces, an analogous result was found independently by Robertson and Walker: there are just three possible isotropic and homogeneous three-dimensional space geometries. The space metric describing these geometries can be given in the form

$$dl^2 = (1 + \tfrac{1}{4}Kr^2)^{-2}(dx^2 + dy^2 + dz^2),$$ (5.8)

where K is a constant and $r^2 = x^2 + y^2 + z^2$. For ease of notation we define $k = K/|K| = \pm 1$ for $K \neq 0$, and $k = 0$ if $K = 0$. If $K \neq 0$, we can put $\bar{x} = |K|^{1/2}x, \bar{y} = |K|^{1/2}y, \bar{z} = |K|^{1/2}z$ to get

$$dl^2 = |K|^{-1}(1 + \tfrac{1}{4}k\bar{r}^2)^{-2}(d\bar{x}^2 + d\bar{y}^2 + d\bar{z}^2),$$ (5.9)

where $\bar{r}^2 = \bar{x}^2 + \bar{y}^2 + \bar{z}^2$. From this we can see that $|K|$ determines the length scale: the two-dimensional analogue is the curvature of the sphere, which determines its size. The quantity $k = \pm 1, 0$ determines the type of geometry, so we have indeed three distinct possibilities here.

Note next that the metric (5.8) satisfies the condition of isotropy about the origin $r = 0$, since the metric coefficients depend only on r^2. The proof of homogeneity or, equivalently, isotropy about any other point is not so easy. Indeed a casual glance at the metric gives the impression that $r = 0$ is a privileged centre, except for the case $K = 0$ where we already know that the space is homogeneous because it is Euclidean. The simplest proof of homogeneity is obtained by considering the space from the point of view of an embedding in a higher-dimensional space.

5.4.1   Homogeneity of the 2-sphere

To see how this works consider again the two-dimensional sphere of unit radius. This is defined by the intrinsic metric

$$ds^2 = d\theta^2 + \sin^2\theta \, d\phi^2,$$ (5.10)

where the angle ϕ denotes longitude and the angle θ denotes colatitude. The metric appears not to be homogeneous since the metric coefficients depend on the coordinate θ; for example the value $\theta = \pi/2$ appears to be privileged. However, consider the surface $x^2 + y^2 + z^2 = 1$ in three-dimensional Euclidean space with coordinates (x, y, z). We can parametrize this surface in terms of coordinates θ and ϕ by writing

$$
\begin{aligned}
x &= \sin\theta \cos\phi \\
y &= \sin\theta \sin\phi \\
z &= \cos\theta
\end{aligned}
$$ (5.11)

since, by substitution, these forms for (x, y, z) certainly satisfy $x^2 + y^2 + z^2 = 1$. What about the intrinsic distances on this surface? Using (5.11), we have

$$dx = \cos\theta \cos\phi \, d\theta - \sin\theta \sin\phi \, d\phi,$$

with similar expressions for dy and dz and, hence, on this surface,

$$dl^2 = dx^2 + dy^2 + dz^2 = d\theta^2 + \sin^2\theta \, d\phi^2.$$

The usual round object in Euclidean space is therefore precisely the surface defined by (5.10) without any reference to the surrounding space.

Now, a rotation in the three-dimensional Euclidean embedding space about the origin of the sphere takes any point on the surface of the sphere into any other. Therefore all points on the surface are equivalent and the sphere is homogeneous. This way of looking at the problem avoids an explicit calculation of the effect of a rotation on points of the sphere in terms of the θ, ϕ coordinates in order to demonstrate that this leaves (5.10) unchanged in form.

5.4.2 Homogeneity of the metric

For the metric (5.8) we make the following parametrization:

$$X = \frac{x}{1 + \frac{1}{4}Kr^2} \qquad Y = \frac{y}{1 + \frac{1}{4}Kr^2}$$

$$Z = \frac{z}{1 + \frac{1}{4}Kr^2} \qquad W = K^{-1/2}\frac{1 - \frac{1}{4}Kr^2}{1 + \frac{1}{4}Kr^2}, \tag{5.12}$$

where $r^2 = x^2 + y^2 + z^2$. We can see straightforwardly that

$$X^2 + Y^2 + Z^2 + W^2 = 1, \tag{5.13}$$

and, after some algebra, that

$$dl^2 = dX^2 + dY^2 + dZ^2 + dW^2 = (1 + \tfrac{1}{4}Kr^2)^{-2}(dx^2 + dy^2 + dz^2).$$

It follows that the geometry described by (5.8) can be thought of as a three-dimensional spherical surface defined by (5.13) in four-dimensional Euclidean space (X, Y, Z, W). Therefore a rotation in the embedding space takes any point of the surface into any other. This demonstrates the homogeneity.

Note that while this description is mathematically correct as an analytic demonstration, it does not provide a correct physical picture of the geometry in all cases. If $k = -1$, the W coordinate is imaginary and the metric of the four-dimensional embedding space is actually Lorentzian, not Euclidean. In this case, therefore, the surface is 'really' a hyperboloid in Minkowski space and the rotation is 'really' a Lorentz transformation.

5.4.3 Uniqueness of the space metric

We have shown that the three space metrics (5.8) are homogeneous and isotropic, but are they the only such possibilities? This is not an easy question. In particular, it is not the same as showing that there are only three types of homogeneous and isotropic three-dimensional surfaces in four-dimensional space, since there might be other possibilities which cannot be embedded in four dimensions. In fact, as Robertson and Walker showed, no such further cases arise. It is important to recall from section 5.2 that this leaves open the possibility of expressing the same metric in many other equivalent ways by using other coordinate systems. We shall, in fact, make use of this freedom in the following sections.

5.4.4 Uniqueness of the spacetime metric

We can use this result of Robertson and Walker to construct the most general four-dimensional spacetime geometry having homogeneous and isotropic spatial sections of constant cosmic time. This is now straightforward. The metric cannot contain terms involving the product $dx\, dt$ since these would change sign under $x \longrightarrow -x$, thereby violating isotropy. We therefore have

$$ds^2 = c^2 f^2(T)\, dT^2 - |K(T)|^{-1}(1 + \tfrac{1}{4}k\rho^2)^{-2}(dx^2 + dy^2 + dz^2), \qquad (5.14)$$

where the spatial part of the metric is (5.9) with the bars dropped, $\rho = (x^2 + y^2 + z^2)^{1/2}$ has been used instead of r, K is an arbitrary function of the time coordinate, T, and f is another arbitrary function of T. A transformation of the time coordinate

$$t = \int f(T)\, dT,$$

so that $dt = f(T)\, dT$, and a relabelling of symbols, $|K(T)|^{-1} = R^2(t)$, takes the metric to the final form

$$ds^2 = c^2\, dt^2 - R^2(t)(1 + \tfrac{1}{4}k\rho^2)^{-2}(dx^2 + dy^2 + dz^2), \qquad (5.15)$$

where, as before, $k = 0, \pm 1$. The function $R(t)$ is the scale factor which we encountered previously. At this stage it is undetermined. It will be given later by the Einstein equations which control how the matter in the Universe determines the geometry (section 5.11).

5.5 Other forms of the metric

Recall that the measured proper time between two events is an invariant, but the expression for the proper time in terms of the coordinates depends on the choice of coordinates. The choice that gives rise to the form (5.15) is not necessarily the most useful one for general calculations, and other coordinates are often employed.

5.5.1   A radial coordinate related to area

The transformation

$$r = \frac{\rho}{(1 + \frac{1}{4}k\rho^2)}, \qquad x = \rho\sin\theta\cos\phi, \qquad y = \rho\sin\theta\sin\phi, \qquad z = \rho\cos\theta$$

in (5.15) leads to

$$ds^2 = c^2\,dt^2 - R^2(t)\left\{ \frac{dr^2}{(1-kr^2)} + r^2(d\theta^2 + \sin^2\theta\,d\phi^2) \right\}. \qquad (5.16)$$

From this form we can see that at time t the area of a sphere of coordinate 'radius' r centred on the origin is $4\pi r^2 R^2(t)$, since on the sphere $r = $ constant ($dr = 0$) and $t = $ constant ($dt = 0$); this is just the Euclidean expression for the area of a sphere of radius $rR(t)$. Thus $rR(t)$ has a physical significance as a measure of area. Note, however, that the spatial geometry is not Euclidean, and $rR(t)$ is not the radius of the sphere as measured by a rigid ruler. We can read off the radial distance between points with coordinates (r, θ, ϕ) and $(r+dr, \theta, \phi)$ at a given time from the metric. It is $dl = R(t)(1 - kr^2)^{-1/2}\,dr$. So the proper radius of the sphere, centred on the origin, $r = 0$, and having radial coordinate r_1 is

$$l = R(t) \int_0^{r_1} (1 - kr^2)^{-1/2}\,dr$$

$$= \begin{cases} R(t)\sin^{-1}r_1 & \text{for } k = 1 \\ R(t)r_1 & \text{for } k = 0 \\ R(t)\sinh^{-1}r_1 & \text{for } k = -1. \end{cases}$$

Thus, l is greater than $R(t)r_1$ for $k = +1$, when the geometry is spherical, l is less than $R(t)r_1$ for $k = -1$, when the geometry is hyperbolic and for $k = 0$ the geometry is Euclidean.

5.5.2   A radial coordinate related to proper distance

Alternatively, we can define a new radial coordinate, χ, such that $\chi R(t)$ is the radial distance. In this case the area of a sphere is not given by $4\pi\chi^2 R^2(t)$. To achieve this put

$$d\chi = \frac{dr}{(1-kr^2)^{1/2}}.$$

Integrating and solving for r gives

$$r = \begin{cases} \sin\chi & \text{for } k = 1 \\ \chi & \text{for } k = 0 \\ \sinh\chi & \text{for } k = -1 \end{cases}$$

so the metric takes the form

$$ds^2 = c^2\,dt^2 - R^2(t)\left\{ d\chi^2 + \left(\frac{\sin^2\sqrt{k}\chi}{\sqrt{k}} \right)(d\theta^2 + \sin^2\theta\,d\phi^2) \right\}. \qquad (5.17)$$

The forms (5.15), (5.16) and (5.17) are those most commonly found in the literature. Henceforth we shall use the metric (5.16).

5.6 Open and closed spaces

In a $k = +1$ Universe with positive curvature the three-dimensional space bends round on itself just as does the positively curved two-dimensional surface of a sphere. Thus we expect the volume of a $k = +1$ Universe to be finite. This can be confirmed by calculating the volume of the model Universe at a fixed time t. The area of a sphere is $4\pi R^2(t)r^2$ and the proper thickness of a spherical shell is $R(t)\,dr/(1 - kr^2)^{1/2}$. So the proper volume, for the case $k = +1$, is

$$V = 2 \times 4\pi R^3 \int_0^{1/\sqrt{k}} \frac{r^2\,dr}{(1 - kr^2)^{1/2}} = 2\pi^2 R^3(t).$$

The integral is taken between one pole ($r = 0$) at the lower limit of integration and $r = 1/\sqrt{k} = 1$ at the 'equator' at the upper limit. Since this covers only half the space we need an additional factor of 2. Since the volume is finite we say that this space is *closed*.

For $k = 0$ and $k = -1$ the integrand is unchanged, but the limits of integration are $r = 0$ to $r = \infty$ in both cases, so the volume is infinite. We say that these spaces are *open*.

5.7 Fundamental (or comoving) observers

Observers located at fixed spatial coordinates in the Robertson–Walker metric are called comoving or fundamental observers. Such observers see the Universe expanding isotropically about them. Our observations of the Universe yield a distribution of matter and an expansion which are both approximately isotropic. To this extent we are fundamental observers. (Of course, 'we' here refers to standard astronomical observers employing standard coordinates, corrections having been made for the motion of the Earth. Since actual observers on the Earth are subject to local gravitational and non-gravitational forces they cannot strictly be fundamental observers.)

Logically the isotropy of the matter distribution and the isotropy of the microwave background are distinct notions. It is possible that the anisotropy of the background radiation which we observe could, in part, be an indication that there are two distinct rest frames. It is, however, difficult to see how such a situation could have evolved, so most cosmologists attribute the observed dipole anisotropy to our peculiar motion relative to an intrinsically isotropic background. In any case, the background temperature deviates from isotropy by only 3×10^{-3}, whether entirely due to our peculiar velocity or not. The smallness of this deviation means again that we are, to a good approximation, fundamental observers.

We do not see any large departures from the Hubble flow, so we conclude that any other observer stationary with respect to their local matter can be regarded as a fundamental observer. For this reason the spatial coordinates, x, are referred to as comoving coordinates because they are anchored to the averaged matter distribution and expand with it. Thus for a fundamental observer

$$x = \text{constant}.$$

Substituting $x = $ constant into any of the metrics forms (5.11) to (5.16) gives

$$d\tau^2 = dt^2,$$

where τ is a proper time. Thus the time t is the proper time kept by fundamental observers: we call this time *cosmic time*. From the cosmological principle all fundamental observers are equivalent; consequently their clocks will run at the same rate. So if we suppose that at $t = 0$ fundamental clocks are set to zero then at any subsequent instant of cosmic time all fundamental observers will see the Universe at the same stage of its evolution with the same mean mass density and the same expansion rate.

There is an important dynamical consistency condition that goes missing in this heuristic approach. We have not shown that the paths in spacetime, $x = $ constant, that we have taken our fundamental observers to follow are, in fact, possible paths in accordance with relativity theory. The result can be proved in the theory, but is in fact clear from symmetry: the isotropy of the metric means that there is no preferred direction in which free falls could depart from the paths of fundamental observers.

5.8 Redshift

We come now to a crucial test. No-one has ever measured an actual change in the distance between galaxies. What we actually refer to as an expansion is deduced from a redshift. So far though, our discussion has been all about changing distances and nothing about redshifts. But in relativity the paths of light-rays, and the wavelengths, are completely determined once the metric distances are given. So there is no room for manoeuvre: either the Robertson–Walker metric gives the Lemaître redshift rule or the theory is wrong.

Consider the emission of successive wavecrests of light from a galaxy G at times t_e and $t_e + dt_e$ (figure 5.2). These are received at times t_0 and $t_0 + dt_0$ respectively by an observer O, located in another galaxy, situated at the origin of the r, θ, ϕ coordinate system. Note that both O and G have constant comoving coordinates. The object of the calculation is to relate the time intervals dt_e and dt_0 at emission and reception and hence the frequencies of the light.

As the observer O is at the origin of coordinates, incoming light-rays move radially, by symmetry and so $\theta = \phi = $ constant. We can therefore put

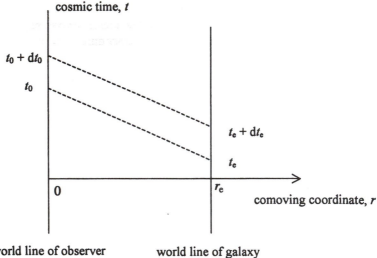

Figure 5.2. World lines of emitter and observer at constant comoving coordinates and the light-rays between them.

$d\theta = d\phi = 0$ along a light-ray approaching the origin. Inserting these values, together with the condition for a light-ray $ds^2 = 0$, into the metric (5.16) gives

$$c\frac{dt}{R} = -\frac{dr}{(1 - kr^2)^{1/2}}, \tag{5.18}$$

where, on taking the square root, the minus sign has been chosen as the ray is moving towards the origin (r decreases as t increases). Integrating along the null-ray from the emission event at (t_e, r_e), to the reception event at $r = 0$ at time t_0, gives

$$c\int_{t_e}^{t_0} \frac{dt}{R} = -\int_{r_e}^{0} \frac{dr}{(1 - kr^2)^{1/2}}. \tag{5.19}$$

Similarly integrating along the second null-ray from the event at $t_e + dt_e$ to the event at $t_0 + dt_0$ gives

$$c\int_{t_e+dt_e}^{t_0+dt_0} \frac{dt}{R} = -\int_{r_e}^{0} \frac{dr}{(1 - kr^2)^{1/2}}.$$

As the observer O and the galaxy G have constant spatial coordinates r, θ, ϕ the right-hand sides of (5.18) and (5.19) are identical. So we may equate their left-hand sides to get

$$c\int_{t_0}^{t_0+dt_0} \frac{dt}{R} = c\int_{t_e}^{t_e+dt_e} \frac{dt}{R}.$$

Since dt_e and dt_0 are negligible compared to $t_0 - t_e$, we have

$$\frac{dt_e}{R(t_e)} = \frac{dt_0}{R(t_0)}. \qquad (5.20)$$

Since dt_e and dt_0 are periods of a wave, $dt_0/dt_e = \nu_e/\nu_0$, where ν_e and ν_0 are the corresponding frequencies of the wave. Hence (5.20) together with the definition of redshift gives

$$1 + z = \frac{\lambda_0}{\lambda_e} = \frac{\nu_e}{\nu_0} = \frac{dt_0}{dt_e} = \frac{R(t_0)}{R(t_e)}, \qquad (5.21)$$

which is the Lemaître redshift rule that we obtained in section 4.4.1. This is how relativity brings redshift and expansion together.

Note that, $\lambda_0/\lambda_e = R(t_0)/R(t_e)$, so the wavelength is proportional to the scale factor which is intuitively comfortable. Therefore, the redshift z of a distant galaxy tells us the amount by which the scale factor has expanded since the light was emitted. The special relativistic Doppler formula for a receding source,

$$1 + z = (1 + v/c)^{1/2}/(1 - v/c)^{1/2}, \qquad (5.22)$$

is *not*, in general, applicable to cosmological redshifts. However, in the limit when $v/c \ll 1$ the Doppler formula (5.22) gives, to a good approximation, $v = cz$ (problem 7). For small z the Lemaître redshift rule together with the velocity–distance law also give $v = cz$ (problem 24). This justifies our use of $v = cz$ in section 2.2.

5.9 The velocity–distance law

We guessed the form of the Robertson–Walker metric on the grounds that it provided for the expansion of distances. But does it provide the right expansion law? In this section we demonstrate that the Robertson–Walker metric gives the relationship between velocity of recession and distance, equation (2.5), that we obtained in chapter 2.

The proper distance l at cosmic time t between the observer O and the galaxy G is, from the metric (5.16),

$$l = \int (-ds^2)^{1/2} = R(t) \int \frac{dr}{(1 - kr^2)^{1/2}} = R(t)f(r). \qquad (5.23)$$

Differentiating with respect to cosmic time t gives

$$\frac{dl}{dt} = \frac{dR}{dt}f(r), \qquad (5.24)$$

where $f(r)$ is not a function of time as galaxies have constant comoving coordinates. So substituting for $f(r)$ from (5.23) into (5.24) gives

$$\frac{dl}{dt} = \frac{\dot{R}}{R}l = Hl,$$

where H is the Hubble parameter.   This is the velocity–distance law, equation (2.5). As we have just shown, it is valid for all distances, and should not be confused with Hubble's law (2.7) which is limited to small redshifts.

5.10   Time dilation

There is one further deduction that we can make from equation (5.20). Let dt_e be the time interval between a pair of events on a galaxy world line and let these events be observed from another galaxy. The relation (5.20) tells us that the observed time interval dt_0 is longer than the time interval dt_e measured at the location of the events by a factor $(1 + z)$.   So the distant observer sees physical processes slowed down by a factor $(1 + z)$. This cosmological time dilation provides us with a test of the expanding Universe picture based on the Robertson–Walker metric. The effect has been observed in the rate of brightening of a supernova at a redshift of $z = 0.86$ (Perlmutter *et al* 1998). Its light curve— the luminosity as a function of time—is indeed stretched by a factor 1.86 relative to equivalent local supernovae, as predicted.

5.11   The field equations

We have seen that the Robertson–Walker metric is obtained from considerations of symmetry alone, without any reference to a specific theory of gravity. But a theory of gravity is needed in order to evaluate the undetermined function of time $R(t)$ which appears in the metric: this theory is Einstein's general theory of relativity.

The theory provides a relationship between the geometry of spacetime and the matter content. As far as the Robertson–Walker metric is concerned the spacetime geometry is contained in the time dependence of the scale function $R(t)$. What about the matter content? We assume that the Universe can be treated at all times as a fluid (gas) of particles described macroscopically by a density ρ and pressure p. These particles might be thought of as the galaxies, or clusters of galaxies, as well as photons and other background matter.

Einstein's field equations then yield two independent relations. First

$$\left(\frac{dR}{dt}\right)^2 = \frac{8}{3}\pi G\rho R^2 - kc^2 + \frac{1}{3}\Lambda R^2, \tag{5.25}$$

which is known as the *Friedmann equation*.

Second

$$\frac{d(\rho R^3)}{dt} + \frac{p}{c^2}\frac{dR^3}{dt} = 0, \tag{5.26}$$

which is the equation of *local energy conservation*.

Differentiating (5.25) with respect to t and using (5.26) gives us the useful equation

$$\frac{d^2 R}{dt^2} = -\frac{4}{3}\pi G R(\rho + 3p/c^2) + \frac{1}{3}\Lambda R, \tag{5.27}$$

which we shall refer to as the *acceleration equation*.

Equation (5.27) is not independent of (5.25) and (5.26) and any pair of the three equations can be taken as the fundamental dynamical equations of cosmology (see problem 27). The cosmological models which are based on the Robertson–Walker metric and the Einstein field equations are often referred to as the Friedmann–Lemaître–Robertson–Walker models or FLRW models (or sometimes just FRW models) for short.

Note that the mean mass density $\rho(t)$ includes contributions from matter in all its various guises and from radiation, which includes background neutrinos if these are massless or relativistic, so $\rho = \rho_m + \rho_r$.

5.11.1 Equations of state

The pressure $p(t)$ arises from the random motion of particles. It enters the field equations in addition to the mass equivalent of the internal energy of motion which is included in ρ. (So ρ is not just the rest mass density.) The pressure appears in the field equations because stresses act as sources of the gravitational field in general relativity. One might imagine that the pressure should help to drive the expansion. However, since the pressure is uniform there are no pressure gradients to give rise to forces. Equation (5.27) shows that, in fact, the effect of a uniform pressure is to *decelerate* the expansion, not to accelerate it.

Since we have only two independent equations for the three unknown functions, $R(t)$, $\rho(t)$ and $p(t)$, the equations would appear not to determine the evolution of the scale factor $R(t)$. However, for all normal matter, p and ρ are not independent. Their relation is known as an equation of state,

$$p = p(\rho),$$

the form of which depends on the nature of the matter under consideration. So, given an equation of state, Einstein's field equations determine $R(t)$.

5.11.2 The cosmological constant

The cosmological constant Λ was originally introduced into the field equations by Einstein as a mathematical device to permit a static Universe to be a solution (problem 26). One way to regard the cosmological term is to think of it as modifying the geometry in response to a given distribution of matter. From this point of view it is interpreted as a curvature of empty spacetime. However, an alternative point of view is to treat it as a source of gravity and to emphasize this by incorporating it into the matter term as an addition to the density. Following this view we now write the Friedmann equation (5.25) as

$$\left(\frac{dR}{dt}\right)^2 = \frac{8}{3}\pi G(\rho + \rho_\Lambda)R^2 - kc^2,$$

where $\rho_\Lambda = \Lambda/(8\pi G)$ is a constant mass density. We consider only the case of positive Λ so the corresponding density ρ_Λ is positive. This appears to be the sign chosen by nature as we shall see in chapter 6.

Looking back to the equation of local energy conservation (5.26) for a constant density ρ_Λ and pressure p_Λ, we have

$$\rho_\Lambda\frac{dR^3}{dt} + \frac{p_\Lambda}{c^2}\frac{dR^3}{dt} = 0.$$

Thus we derive an equation of state for this 'matter':

$$p_\Lambda = -\rho_\Lambda c^2. \tag{5.28}$$

We shall see in chapter 8 that the ground state of a quantum field has just these properties of positive mass density and negative pressure. This provides a natural physical interpretation of the cosmological constant.

Applying our reinterpretation to the acceleration equation (5.27) we rewrite this as

$$\frac{d^2R}{dt^2} = -\frac{4}{3}\pi GR[\rho + \rho_\Lambda + 3(p + p_\Lambda)/c^2], \tag{5.29}$$

from which we can return to the original form by substituting for ρ_Λ and p_Λ from (5.28).

In the absence of normal matter, $\rho = p = 0$ and the only source of gravity is from the density and pressure of the cosmological term. Equation (5.29) becomes

$$\frac{d^2R}{dt^2} = \frac{8}{3}\pi GR\rho_\Lambda.$$

Note that the negative pressure has made d^2R/dt^2 positive. This makes it clear that a negative pressure fluid gives rise to repulsive gravitation. It is worth emphasizing this point as it can be a source of confusion. A positive cosmological constant is equivalent to a positive mass density, so ρ_Λ adds to the other mass densities. But because of its curious equation of state a negative-pressure cosmic fluid gravitates in the opposite sense to other sources of mass.

5.11.3 The critical density

In chapter 3 we introduced the critical density $\rho_c = 3H^2/8\pi G$ as a convenient scaling parameter. We can now investigate its physical significance. Dividing the

Friedmann equation (5.25) through by R^2 and rearranging the order of the terms gives

$$-kc^2 = R^2 H^2 \left(1 - \frac{8\pi G\rho}{3H^2} - \frac{\Lambda}{3H^2} \right), \tag{5.30}$$

where we have used $H = \dot{R}/R$. The density ρ here includes all contributions other than that from the cosmological term, so it includes both matter and radiation. This equation shows how the spatial curvature k is related to the density ρ of matter and radiation and the density $\rho_\Lambda = \Lambda/(8\pi G)$ associated with the cosmological constant Λ. The spatial curvature is zero, that is $k = 0$, if

$$\rho + \rho_\Lambda = \frac{3H^2}{8\pi G}. \tag{5.31}$$

Thus a critical density Universe has zero spatial curvature.

A Universe with negative spatial curvature has $k = -1$ and so, from (5.30),

$$\rho + \rho_\Lambda < \frac{3H^2}{8\pi G}.$$

Thus the total density is less than the critical value. Similarly a positively curved Universe ($k = +1$) has a total density greater than the critical value.

This relation between mass density and geometry revealed here is an example of the connection that exists in the general theory of relativity between the spacetime metric and the mass distribution: in general each determines the other in a self-consistent manner.

The density parameter Ω_m, which we met in section 3.1, expresses the matter density in terms of the critical density

$$\Omega_m \equiv \frac{\rho_m}{\rho_c} = \frac{8\pi G\rho_m}{3H^2}, \tag{5.32}$$

with a similar expression for the radiation

$$\Omega_r = \frac{\rho_r}{\rho_c} = \frac{8\pi G\rho_r}{3H^2},$$

and the cosmological term,

$$\Omega_\lambda = \frac{\rho_\Lambda}{\rho_c} = \frac{8\pi G\rho_\Lambda}{3H^2}.$$

(Recall our convention uses Ω_λ to denote the time-dependent value, Ω_Λ the value at the present time; the equivalent cosmological density ρ_Λ is constant so there is no distinction.) Using the density parameters we can write the Friedmann equation (5.30) in the neater form

$$-kc^2 = R^2 H^2 (1 - \Omega_m - \Omega_r - \Omega_\lambda). \tag{5.33}$$

We see that the total density parameter $\Omega_{\text{tot}} = \Omega_m + \Omega_r + \Omega_\lambda$ determines the spatial curvature as follows:

$$\Omega_{\text{tot}} = 1 \text{ implies } k = 0$$
$$\Omega_{\text{tot}} < 1 \text{ implies } k = -1$$
$$\Omega_{\text{tot}} > 1 \text{ implies } k = +1.$$

5.12   The dust Universe

The mass density of matter at the present time is

$$\rho_M = 1.88 \times 10^{-26} h^2 \Omega_M \text{ kg m}^{-3}$$

(equation (3.5)). Taking $h = 0.65$ and $\Omega_M = 0.3$ we see that the mass density in the form of matter, both dark and baryonic, is currently about 2×10^{-27} kg m^{-3}. By comparison, the current mass density of the microwave background radiation is 4.6×10^{-31} kg m^{-3} (problem 17) and the mass density of the background neutrinos, if these have zero rest mass, should be 0.68 times the mass density of the photons (see section 7.2). If neutrinos have non-zero rest masses less than 1 eV/c^2, as appears likely at the present time, their contribution to ρ_0 is still very small. Consequently, if we ignore for now the possible contribution from ρ_Λ, we have $\rho_0 \simeq \rho_M$ to a good approximation at the present time. Under these conditions the Universe is said to be *matter dominated*. Moreover, the pressure of this material is close to zero. The intergalactic medium has negligible pressure because it is extremely tenuous; the pressure corresponding to the random motions of the galaxies and clusters of galaxies is negligible because these motions are small (problem 28).

To a good approximation the equation of state of the cosmic gas of galaxies is, therefore,

$$p = 0. \tag{5.34}$$

Cosmological tradition has it that such a pressureless non-viscous fluid is called 'dust' for reasons that are now obscure. In the dust model of the Universe it is assumed that $p = 0$ at all times. In fact, this approximation is a good one for most of the history of the Universe as we shall show in section 5.16. We shall develop a model appropriate to the early history, when the dust model does not apply, in the next section.

Inserting the equation of state (5.34) into the equation of local energy conservation (5.26) gives

$$\rho R^3 = \text{constant}. \tag{5.35}$$

This is just a statement of the conservation of mass.

For the present we shall confine ourselves to the case $\Lambda = 0$ and consider what happens when $\Lambda \neq 0$ in section 5.15. With this assumption the Friedmann equation reduces to

$$\left(\frac{dR}{dt}\right)^2 = \frac{8}{3}\pi G\rho R^2 - kc^2. \tag{5.36}$$

Since the equation of state eliminates p, the two equations (5.35) and (5.36) contain only two unknown functions, so we can solve them for $R(t)$. We shall derive some general results first and then treat the three solutions corresponding to the three values of k in turn.

5.12.1 Evolution of the density parameter

In every FLRW model there is a unique relation between the observed redshift of a galaxy and the cosmic time at which the light must have been emitted in order to be visible to the observer. Consequently, we can refer to events occurring at redshift z instead of at time t. Often the evolution of a physical quantity takes a simpler form when expressed in terms of redshift and, in any case, such an expression is more closely related to observation since we deduce times from measured redshifts.

In dust models the conservation of mass as a function of time, $\rho R^3 = $ constant, can be written immediately as a relation between density and redshift. Using the Lemaître redshift rule (5.21),

$$1 + z = \frac{R_0}{R}$$

we get

$$\rho = \rho_0(1+z)^3.$$

We now seek a relation that will tell us how the density parameter $\Omega = 8\pi G\rho/3H^2$ behaves as a function of redshift. Since $\Omega \propto \rho/H^2$ we get

$$\frac{\Omega}{\Omega_0} = \frac{\rho}{\rho_0}\frac{H_0^2}{H^2} = (1+z)^3\frac{H_0^2}{H^2}$$

or

$$\Omega H^2 = (1+z)^3\Omega_0 H_0^2. \tag{5.37}$$

To complete the analysis we need an expression for the evolution of the Hubble parameter H in terms of z.

5.12.2 Evolution of the Hubble parameter

The Friedmann equation in the form (5.33) with $\Omega_\Lambda = 0$ gives

$$-kc^2 = R^2 H^2(1 - \Omega) = R_0^2 H_0^2(1 - \Omega_0), \tag{5.38}$$

where the second equality follows because $-kc^2$ is not a function of time. Hence

$$H^2(1 - \Omega) = \frac{R_0^2}{R^2}H_0^2(1 - \Omega_0) = (1+z)^2 H_0^2(1 - \Omega_0). \tag{5.39}$$

Now (5.37) and (5.39) are two equations for H and Ω. Substituting for $H^2\Omega$ from (5.37) into (5.39) gives

$$H = H_0(1+z)(1+\Omega_0 z)^{1/2} \tag{5.40}$$

as the evolution equation for the Hubble parameter, and then (5.37) gives

$$\Omega = \frac{\Omega_0(1+z)}{(1+\Omega_0 z)}$$

as the evolution equation for the density parameter in pressure-free models with zero cosmological constant.

5.13   The relationship between redshift and time

We can now complete the picture by obtaining the relation between t and z in the dust models. We will derive it here for the case $\Lambda = 0$. We have, from the Lemaître redshift rule (5.21),

$$1 + z = R_0/R$$

and hence, differentiating,

$$\frac{dz}{dt} = -\left(\frac{R_0}{R^2}\right)\frac{dR}{dt} = -(1+z)\frac{1}{R}\frac{dR}{dt} = -(1+z)H. \tag{5.41}$$

The object of the exercise is to write this relation in terms of the measurable parameters H_0 and Ω_0. We can do this immediately by using the evolution equation for the Hubble parameter (5.40) to get the required differential equation

$$\frac{dz}{dt} = -H_0(1+z)^2(1+\Omega_0 z)^{1/2}. \tag{5.42}$$

Note that this equation is valid for all values of k. Since this was obtained from the Friedmann equation, which itself follows from Einstein's equations for the Robertson–Walker metric, we can regard (5.42) as an alternative form of one of the Einstein equations, but now telling us how the geometry explored by light-rays is influenced by matter. For the case $\Lambda \neq 0$ see chapter 6 (section 6.2.2) and problem 43.

A dust Universe with the critical density, $\Omega = 1$, is called the Einstein–de Sitter model (see later). In this case, equation (5.42) integrates to give

$$t = \frac{t_0}{(1+z)^{3/2}}. \tag{5.43}$$

The results of integrating (5.42) for different values of Ω_0 are plotted in figure 5.3.

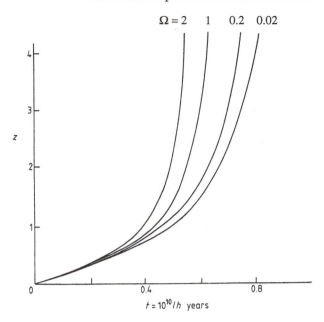

Figure 5.3. A plot of redshift z against time t for different values of Ω. The asymptotes $z \to \infty$ give the ages for the different models.

5.13.1 Newtonian interpretation

The Friedmann equation (5.36) can be interpreted in terms of a simple Newtonian picture. Rearranging the equation slightly, we can write

$$\frac{1}{2}\left(\frac{dR}{dt}\right)^2 - G\frac{(\frac{4}{3}\pi\rho_0 R_0^3)}{R} = -\frac{1}{2}kc^2. \tag{5.44}$$

We can interpret this as an energy equation. In Newtonian language $\frac{1}{2}(dR/dt)^2$ is the kinetic energy per unit mass of a particle with velocity dR/dt; the second term is the potential energy per unit mass due to the sphere of matter interior to the particle at radius R; and the third term is the constant total energy. The equation as a whole then states that kinetic energy plus potential energy is conserved. We can therefore deduce the fate of these Universe models from the Newtonian picture. If $k = +1$ the total energy is positive and the unit mass escapes to infinity in a finite time. If $k = 0$, the total energy is zero, so the system is marginally bound: if dR/dt is positive at any time then the system just escapes to infinity in infinite time. If $k < 0$ the total energy is negative: so the system is bound and even if it is expanding at any one time it must eventually collapse. These deductions are rigorous because, as we shall show later, they depend only on the mathematical form of the equation. The interpretation in terms of Newtonian energy simply allows us to bring our physical intuition to bear on the form of the solutions. What is not so clear is the extent to which the Newtonian picture has any deeper significance, in particular whether a Newtonian derivation of (5.44) is meaningful or a fortuitous working back from the right answer. Some authors make a plausible case that equation (5.44) can be derived exactly even for relativistic expansion speeds. The reader may consider it odd that an exact relativistic equation for the field can be *deduced* from the non-relativistic Newtonian equation of motion for a particle.

5.14 Explicit solutions

In the following sections we shall find the explicit behaviour of the scale factor $R(t)$ as a function of time in some special cases.

5.14.1 $p = 0, k = 0, \Lambda = 0$, the Einstein–de Sitter model

The zero-pressure model with zero spatial curvature, $k = 0$ is known as the Einstein–de Sitter model. The Friedmann equation for this case can be written

$$\left(\frac{dR}{dt}\right)^2 = \frac{8}{3}\pi G\rho R^2. \tag{5.45}$$

Using conservation of mass, $\rho R^3 = $ constant, equation (5.35), to eliminate ρ, equation (5.45) is easily integrated to give

$$R(t) = (6\pi G\rho_0 R_0^3)^{1/3} t^{2/3}, \tag{5.46}$$

in which the constant of integration has been chosen so that, at $t = 0$, $R(0) = 0$. Note that, whatever the constant of integration, $R(t)$ must become zero at some finite time in the past and it makes no difference to the physics that we conventionally measure time from this point.

Since $\rho R^3 = $ constant we can put $\rho R^3 = \rho_0 R_0^3$ in (5.46) to obtain

$$\rho(t) = (6\pi Gt^2)^{-1}$$

for the density as a function of time. Evidently at $t = 0$, the density is infinite and we say there is a 'singularity'. In addition, the redshift $z = (R_0/R) - 1$, becomes infinite, so we shall not receive any radiation from this time. Thus physics cannot tell us what happens at $t = 0$, or before, and this model begins abruptly with the coming into existence of space and time, an event that is referred to as the big bang.

Equation (5.46) tells us that, subsequent to $t = 0$, $R(t)$ increases monotonically with time. The galaxies move apart but the rate of expansion decreases and tends to zero at large times (figure 5.4). In chapter 2 we estimated

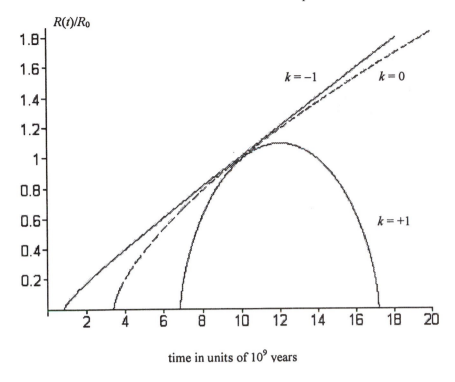

time in units of 10^9 years

Figure 5.4. The evolution of the scale factor for three models having $\Lambda = 0$ and the same value of the Hubble constant at the present time. The curves are labelled with the space curvature.

the age of the Universe to be somewhat less than H_0^{-1}. Now we can be more precise. The age of an Einstein–de Sitter Universe is obtained from the relation

$$H = \frac{\dot{R}}{R}.$$

Since $R(t) \propto t^{2/3}$, equation (5.46), we get $H = 2/3t$. Thus the present age, t_0, is given by

$$t_0 = \frac{2}{3H_0} = 6.52 \times 10^9 h^{-1} \text{ years}, \tag{5.47}$$

which, for $h = 0.65$, is 1.0×10^{10} years. Of course, whilst this is exact for the Einstein–de Sitter model, it is not necessarily a good estimate of the age of the real Universe. We discuss the age of the Universe in greater depth in chapter 6.

Looking at the Friedmann equation (5.36) we see that the first term on the right-hand side, $\frac{8}{3}\pi G\rho R^3/R = \frac{8}{3}\pi G\rho_0 R_0^3/R$, where we have multiplied top and bottom by R, is inversely proportional to $R(t)$. It will therefore be very large at early times. However, the curvature term $-kc^2$ is independent of time. Thus at sufficiently early times the curvature term must be negligible and the Friedmann equation will take the form (5.45) irrespective of the value of k. So at early times all three models behave like the Einstein–de Sitter model and only later does the curvature term start to manifest itself (problem 29).

5.14.2 The case $p = 0, k = +1, \Lambda = 0$

The Friedmann equation (5.36) with $k = +1$ gives

$$\left(\frac{dR}{dt}\right)^2 = \frac{8}{3}\pi G\rho R^2 - c^2. \tag{5.48}$$

The new feature of this model is that expansion comes to a halt, that is $dR/dt = 0$, when

$$R = R_{max} = \tfrac{8}{3}\pi G\rho_0 R_0^3/c^2, \tag{5.49}$$

where we have used the conservation of mass, $\rho R^3 = \text{constant} = \rho_0 R_0^3$. We use the form (5.33) of the Friedmann equation at the present time to get R_0 in terms of measurable quantities:

$$c^2 = R_0^2 H_0^2 (\Omega_0 - 1). \tag{5.50}$$

It would appear that this formula could be rearranged to give us R_0 in terms of measurable quantities and this would appear to contradict our previous assertion that only a ratio of scale factors, not R_0 itself, is measurable. However, the equation has been derived by arbitrarily putting $k = +1$. Had we kept k explicit, equation (5.50) could be used to obtain only the combination R_0^2/k, with k, and hence R_0 itself, arbitrary. Now, using (5.50), R_{max} is given by

$$R_{max} = \tfrac{8}{3}\pi G\rho_0 R_0^3/c^2 = H_0^2 \Omega_0 R_0^3/c^2 \tag{5.51}$$

$$= \frac{c\Omega_0}{H_0(\Omega_0 - 1)^{3/2}}, \tag{5.52}$$

the final expression involving only measurable quantities. Beyond the maximum of the expansion, the radius of the model contracts to $R = 0$ again. (Since the acceleration equation (5.27) implies $\ddot{R} < 0$ throughout.)

To show that $R \to 0$ in a finite time we need to obtain the full solution for $R(t)$. By using the definition of R_{max}, equation (5.51), we can rewrite the Friedmann equation (5.36) in the neater form

$$\left(\frac{dR}{dt}\right)^2 = c^2\frac{R_{max}}{R} - c^2.$$

Separating the variables gives

$$\int_0^R \frac{R^{1/2}}{(R_{max} - R)^{1/2}}dR = c\int_0^t dt.$$

This may be integrated by making the usual trigonometric substitution

$$R(\eta) = R_{\max} \sin^2(\eta/2), \tag{5.53}$$

where the angle is customarily taken as $\eta/2$. Integration then gives

$$t(\eta) = (R_{\max}/2c)(\eta - \sin \eta). \tag{5.54}$$

Eliminating η, in principle, between (5.53) and (5.54) gives $R(t)$. In practice we cannot eliminate η but nevertheless these equations give a parametric representation, with parameter η, of the R–t plot. The curve is, in fact, a cycloid (figure 5.4). With $\eta = \pi$ equation (5.53) yields $R(\pi) = R_{\max}$ and with $\eta = 2\pi$ equation (5.53) gives $R(2\pi) = 0$. Thus this model starts with a big bang and ends with a 'big crunch' (see figure 5.4).

Let us now obtain the present age t_0 of such a Universe in terms of measurable quantities. We can get η_0 from (5.53) by solving for η and substituting for $R_{\max}$:

$$\eta_0 = 2 \sin^{-1}(1 - \Omega_0^{-1})^{1/2}. \tag{5.55}$$

Then substituting for $R_{\max}$ from (5.52) and for $\eta = \eta_0$ from (5.55) into (5.54) gives

$$t_0 = \frac{\Omega_0}{H_0(\Omega_0 - 1)^{3/2}} \left\{ \sin^{-1}(1 - \Omega_0^{-1})^{1/2} - \frac{(\Omega_0 - 1)^{1/2}}{\Omega_0} \right\}. \tag{5.56}$$

This can be checked by considering the limit $\Omega_0 \to 1$, in which case we should regain the Einstein–de Sitter result (5.47) (problem 30).

For Ω_0 just greater than unity equation (5.56) reduces to $t_0 \simeq 2(3H_0\Omega_0^{1/2})$. For large Ω_0 we see that $t_0 < (2/3)H_0^{-1}$ and the age of the $k = +1$ Universe is appreciably less than the ages of the stars that it is supposed to contain. Indeed, even the Einstein–de Sitter Universe with $h = 0.65$ appears to be ruled out by this consideration as the ages of the oldest stars come out to be about 1.3×10^{10} years.

Note that the expressions for the ages depend only on the two parameters Ω_0 and H_0 or, additionally, on Λ if this is non-zero. This will be true also for other observable relations to be derived later in chapter 6. No further parameters involving higher derivatives of the scale factor are required. This, of course, is a consequence of the field equations by which the higher derivatives are implicitly related to R, dR/dt and d^2R/dt^2 only.

For these $\Lambda = 0$ models we can conclude that if $\Omega_0 > 1$ the Universe must collapse. In these cases the density of matter not only determines the spatial geometry but also the future fate of the model. (This is not true if $\Lambda \neq 0$.) This is easily interpreted in Newtonian terms (section 5.13.1): if the Friedmann equation is regarded as an energy equation, then $k = +1$ implies that the total energy is negative; hence the system is bound and cannot expand to infinity. For a sufficient density of matter there is enough gravitational attraction to overcome the initial expansion and induce collapse. At some time $t = t_f$ we again get a singularity beyond which the theory is powerless to predict.

5.14.3 The case $p = 0$, $k = -1$, $\Lambda = 0$

With $k = -1$ the Friedmann equation (5.36) becomes

$$\left(\frac{dR}{dt}\right)^2 = \frac{8}{3}\pi G\rho R^2 + c^2. \tag{5.57}$$

In this model there is no value of $R(t)$ for which $\dot{R}(t) = 0$: for all $R(t)$ we have $\dot{R} > 0$ and the Universe expands for ever. Near $t = 0$ we can neglect the k term and again $R(t) \propto t^{2/3}$. For large $R(t)$ we conclude, from the argument used in section 5.14.1, that the first term on the right-hand side of the Friedmann equation (5.57) is much smaller than c^2 so we can set $(dR/dt)^2 = c^2$. This integrates to give

$$R(t) = ct + \text{constant}.$$

The behaviour in this case differs from the $k = 0$ model, where $\dot{R}(t) \to 0$ as $t \to \infty$, since here we have shown that $\dot{R}(t)$ is finite at infinite time. This behaviour accords with the Newtonian interpretation of equation (5.36): $k = -1$ means positive total energy and an unbounded system.

We can integrate the Friedmann equation just as we did in the $k = +1$ case to get $R(t)$. Once again R and t are obtained in terms of a parameter. Doing this gives

$$R(\eta) = R_{\max} \sinh^2 \eta/2 \tag{5.58}$$
$$t(\eta) = (R_{\max}/2c)(\sinh\eta - \eta).$$

where here $R_{\max}$ is given by

$$R_{\max} = \frac{8\pi G\rho_0 R_0^3}{3c^2} = \frac{c\Omega_0}{H_0(1 - \Omega_0)^{3/2}},$$

which follows from the Friedmann equation in the form (5.33) and the definition (5.51). Note that in this case $R_{\max}$ is not the maximum radius (there is none) but has a different interpretation: it is the radius of the black hole containing a mass $\frac{4}{3}\pi\rho_0 R_0^3$. A plot of $R(t)$ against t is shown in figure 5.4 with the other two models for comparison.

For the $k = -1$ model a calculation similar to that of the $k = +1$ case gives for the present age

$$t_0 = \frac{\Omega_0}{(1 - \Omega_0)^{3/2}H_0}\left\{\frac{(1 - \Omega_0)^{1/2}}{\Omega_0} - \sinh^{-1}(\Omega_0^{-1} - 1)^{1/2}\right\}. \tag{5.59}$$

For Ω_0 just less than 1 this expression reduces to $t_0 \simeq 2/(3H_0\Omega_0^{1/2})$ in a similar manner to equation (5.56). Thus $(2/3)H_0^{-1} < t_0 < H_0^{-1}$. The upper limit to t_0 corresponds to the limiting case of vanishingly low density . For example,

$\Omega_0 = 0.3$ and $H_0 = 65$ km s^{-1} Mpc^{-1} give $t_0 = 1.2 \times 10^{10}$ years and $\Omega_0 \ll 1$ gives $t_0 = 1/H_0 = 1.5 \times 10^{10}$ years. It is, therefore, possible to accommodate stellar ages with a suitably low value for Ω_0. Figure 5.3 illustrates the influence of Ω_0 on age for the three models having zero cosmological constant.

5.15 Models with a cosmological constant

In this section we will consider how the three dust solutions of the Friedmann equation, with each of the three values of k, are modified by the inclusion of the cosmological term. Our starting point is, therefore, the Friedmann equation in the form

$$\left(\frac{dR}{dt}\right)^2 = \frac{8}{3}\pi G\rho R^2 - kc^2 + \frac{1}{3}\Lambda R^2 \qquad (5.60)$$

and the conservation of mass equation which applies to pressureless matter (dust):

$$\rho R^3 = \text{constant.} \qquad (5.61)$$

We will also need the acceleration equation (5.27) which, with $p = 0$, is

$$\frac{d^2R}{dt^2} = -\frac{4}{3}\pi G R\rho + \frac{1}{3}\Lambda R. \qquad (5.62)$$

First we note the different time dependence of the three terms on the right-hand side of the Friedmann equation (5.60). As we saw in section 5.14.1 the first term is inversely proportional to R, the second is independent of R and now we have a third term which is proportional to R^2. Observational constraints limit the magnitude of the cosmological term at the present time to be, at most, of the order of the first term in the Friedmann equation (see section 6.4.3). Consequently at early times the curvature term and the cosmological term will not come into play and $R(t)$ will approximate closely to the Einstein–de Sitter solution, $R(t) \propto t^{2/3}$. On the other hand, in cases where the expansion continues indefinitely the cosmological term will dominate the other two terms at large times.

5.15.1 Negative Λ

Consider first the case $\Lambda < 0$. For $k = +1$ the Λ term has the same sign as the curvature term and thus acts with it to halt the expansion. For $k = -1$ the same thing happens because the cosmological term will eventually overwhelm the positive curvature term. Clearly the $k = 0$ case suffers the same fate. Furthermore the acceleration equation (5.27) tells us that, d^2R/dt^2 is always negative as the two terms on its right-hand side are both negative. Thus a Universe with $\Lambda < 0$ eventually recollapses irrespective of whether it is open or closed (figure 5.5). So open and closed spaces do not correspond to continued expansion and recollapse respectively, except in special cases, even though, mistakenly, they are popularly believed to do so.

5.15.2 Positive Λ

Now we consider the case of $\Lambda > 0$. For $k = -1$ the cosmological term has the same sign as the curvature term so the expansion continues indefinitely. The same conclusion holds for the $k = 0$ case. For both cases in the limit when $t \gg t_0$ only the cosmological term is significant and the Friedmann equation reduces to

$$\left(\frac{\mathrm{d}R}{\mathrm{d}t}\right)^2 \simeq \frac{1}{3}\Lambda R^2, \tag{5.63}$$

which has the solution

$$R(t) = \text{constant} \times \exp(Ht),$$

where $H = (\Lambda/3)^{1/2}$ is the Hubble constant. A model in which $\rho = 0$ and $\Lambda > 0$ is called a de Sitter Universe. We see from (5.62) that $\mathrm{d}^2 R/\mathrm{d}t^2$ is positive, which means that the expansion is accelerating, that is the deceleration parameter q is negative.

5.15.3 Positive Λ and critical density

Recent measurements (section 6.4.3) imply a negative value for q and hence a positive cosmological constant and a density close to the critical value. So let us now look at such a critical density model. The Friedmann equation with $k = 0$ is

$$\left(\frac{\mathrm{d}R}{\mathrm{d}t}\right)^2 = \frac{8}{3}\pi G\rho_\mathrm{m} R^2 + \frac{1}{3}\Lambda R^2.$$

We can separate the variables and integrate. Doing this gives

$$R(t) = R_0 \left(\frac{\rho_\mathrm{M}}{\rho_\Lambda}\right)^{1/3} \sinh^{2/3}\left[\frac{(3\Lambda)^{1/2}}{2}t\right], \tag{5.64}$$

where $\rho_\Lambda = \Lambda/8\pi G$ and $\rho_\mathrm{M} + \rho_\Lambda = 3H_0^2/8\pi G$. The age of this model is obtained from (5.64) by setting $R = R_0$ and $t = t_0$. After some algebra we get

$$t_0 = \frac{2}{3H_0\Omega_\Lambda^{1/2}} \tanh^{-1}\Omega_\Lambda^{1/2},$$

where $\Omega_\Lambda = \Lambda/3H_0^2$ (see problem 34). So, for example, with $\Omega_\Lambda = 0.7$ and $H_0 = 65$ km s^{-1} Mpc^{-1}, $t_0 = 1.45 \times 10^{10}$ years. Comparing this age with that given in equation (5.47) we see that the inclusion of a positive cosmological constant extents the age of a critical density model. Such a model can accommodate the ages of the oldest stars. A plot of $R(t)$ for this critical density case is shown in figure 5.5.

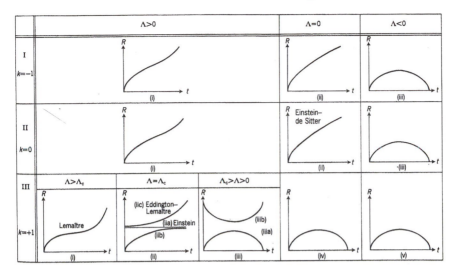

Figure 5.5. The evolution of the scale factor in models with various values of Λ (adapted from D'Inverno 1992).

5.15.4   The case $\Lambda > 0, k = +1$

Finally, we consider $\Lambda > 0$ and $k = +1$. Recall from section 5.14.2 that when $\Lambda = 0$ and $k = +1$ expansion is momentarily brought to a halt at $R_{\max}$ given by (5.51), $R_{\max} = c\Omega_0 H_0^{-1}(\Omega_0 - 1)^{-3/2}$. The acceleration equation (5.27) tells us that collapse will follow because, for $\Lambda = 0$, the second derivative of R is always negative, as ρ is positive for normal matter. Now, with $\Lambda > 0$, the Friedmann equation (5.25) tells us that the expansion is brought to rest when

$$\Lambda = \frac{3c^2}{R^2} - 8\pi G \frac{\rho_0 R_0^3}{R^3},$$

where we have used the constancy of ρR^3 to substitute present values. This function $\Lambda(R)$ has a maximum at $R = R_c = 4\pi G(\rho R^3)_0/c^2$, at which point $\Lambda = \Lambda_c = c^2/R_c^2$. On substituting these critical values into the acceleration equation (5.27) we see that $\mathrm{d}^2 R/\mathrm{d}t^2 = 0$ at all times. This is the Einstein Universe in which the repulsive effect of the positive cosmological constant just balances the attraction of the gravity due to the matter.

For $\Lambda > \Lambda_c$ the repulsion overcomes gravity and the expansion is not halted. For $\Lambda < \Lambda_c$ the expansion is halted and collapse follows. The full range of solutions are shown in figure 5.5.

5.16   The radiation Universe

As we go back in time the matter density increases as $\rho_m \propto R^{-3}$ or, equivalently, $\rho_m \propto (1+z)^3$. How does the radiation behave? Including a contribution from neutrinos (see chapter 7) we have $\rho_r = 1.68aT^4/c^2$ with $T \propto R^{-1}$, so $\rho_r \propto R^{-4}$ or $\rho_r \propto (1+z)^4$. Hence the radiation density increases more rapidly as we go back in time than the matter density. We can calculate the redshift, z_{eq}, at which the two densities are equal from

$$\rho_r = \rho_R(1+z_{eq})^4 = \rho_M(1+z_{eq})^3 = \rho_m.$$

Therefore

$$1 + z_{eq} = \frac{\rho_M}{\rho_R} = 2.4 \times 10^4 \Omega_M h^2$$

(see problem 58). We have included in ρ_r a contribution, aT^4/c^2, from the background radiation and a contribution, $0.68aT^4/c^2$, from neutrinos which, even if they have a small mass, were relativistic at $z_{eq} \sim 10^4$ (see section 7.2). At times prior to z_{eq} the radiation density was dominant. In the Einstein–de Sitter model, with $h = 0.65$ this redshift corresponds to a time $t_{eq} \simeq 10\,000$ years. Before this time the dust model cannot be appropriate. We say that the early Universe was *radiation dominated*(see section 7.4 for an accurate value of t_{eq}).

At the opposite extreme from a Universe filled only with cold matter is a Universe filled only with radiation. Such a model will be a suitable approximation to the Universe at times prior to z_{eq} when the mass density of radiation, ρ_r, and hence its gravitational effect, substantially exceeds that of matter. This means that the density of the radiation controls the dynamics of the expansion, that is $\rho = \rho_r$ in the field equations.

The radiation gas exerts a pressure. This is given by the equation of state for radiation

$$p = \tfrac{1}{3}\rho_r c^2. \tag{5.65}$$

Substituting this relation into the equation of local energy conservation, (5.26), multiplying by $R(t)$ and setting $\rho = \rho_r$ yields

$$R^4 \frac{d\rho_r}{dt} + 4\rho_r R^3 \frac{dR}{dt} = 0,$$

which integrates to give

$$\rho_r R^4 = \text{constant}. \tag{5.66}$$

This is the same result as the one we obtained by a different argument in the first paragraph of this section. The two are consistent because the preservation of a blackbody spectrum is consistent with local energy conservation.

We have already seen that at early times the curvature and cosmological terms are small and can be neglected. So the Friedmann equation (5.25) for a radiation Universe simplifies to

$$\left(\frac{dR}{dt}\right)^2 = \frac{8}{3}\pi G \rho_r R^2. \tag{5.67}$$

Integrating (5.67) with the help of (5.66) gives

$$\frac{R}{R_0} = \left(\frac{32}{3}\pi G \rho_R\right)^{1/4} t^{1/2}, \tag{5.68}$$

where $\rho_R = \rho_r(t_0)$, which is the equation of a parabola. This is the unique solution for the scale factor at early times, i.e. for $z \gg z_{eq}$, or $t \ll 10\,000$ years.

Finally, we can obtain a relationship between redshift, or $R(t)$, and time for epochs in which the curvature and cosmological terms can be neglected by integrating the Friedmann equation with $\rho = \rho_m + \rho_r$. This gives

$$t = 1.76 \times 10^3 \left[\left(\frac{R}{R_{eq}} - 2\right)\left(\frac{R}{R_{eq}} + 1\right)^{1/2} + 2\right] (\Omega_M h^2)^{-2} \text{ yr}, \tag{5.69}$$

where $R_0/R_{eq} = \rho_R/\rho_M$ (problem 32).

5.16.1 The relation between temperature and time

For a blackbody spectrum $TR = T_0 R_0 = $ constant. Using this in equation (5.68) gives the following important relation between the temperature of the radiation and the time in a radiation Universe:

$$T = \left(\frac{3c^2 T_0^4}{32\pi G u_R}\right)^{1/4} t^{-1/2}. \tag{5.70}$$

Explicitly, with $u_R = 1.68 a T_0^4$

$$T = \frac{1.3 \times 10^{10}}{t^{1/2}} \text{ K}, \tag{5.71}$$

with t in seconds. Thus at an age of 1 s, the radiation Universe has a temperature of approximately 10^{10} K. Note that prior to 1 s the factor of 1.68 in u_R has to be replaced by a different one for reasons that will be explained in chapter 7. The general form of the relation between time and temperature remains valid whenever curvature can be neglected and the density is dominated by radiation and matter in which the internal speeds are close to the speed of light (referred to as *relativistic matter*). For relativistic matter the rest mass can be neglected and, under such circumstances, the mass-density just arises from the internal energy and is determined solely by the temperature.

5.17 Light propagation in an expanding Universe

When we look at an object with a high redshift we are looking a long way into the past. A deep survey of the sky shows galaxies with a wide range of redshifts, so direct observation shows galaxies where they were, and how they were at a range of different times. It does not show us the Universe as it is today, except nearby. We call this view of the Universe a world picture. We would like a view of the Universe at a given time and also to be able to see how it evolves in time. Such a view is called a world map and is realized by plotting the world lines of galaxies and light-rays on a spacetime diagram. Let us see the insights that a world map gives us.

Take our location to be at the origin of the r, θ, ϕ coordinates of the Robertson–Walker metric (5.16) and consider the propagation of light from a distant source towards us. Then, from the symmetry of the metric, incoming light-rays travel radially towards the origin, so $\theta = \phi = $ constant. Inserting this condition, together with the null-ray condition, $ds = 0$, into the metric (5.16) gives

$$c\,dt = -R\,dr\,(1 - kr^2)^{-1/2},$$

where, on taking the square root, the minus sign has been chosen because the ray is moving in the direction of r decreasing.

So for a photon which passes radial coordinate r at time t and reaches us at $r = 0$ and $t = t_0$ we have

$$\int_r^0 \frac{dr}{(1 - kr^2)^{1/2}} = -c \int_t^{t_0} \frac{dt}{R}. \tag{5.72}$$

Equation (5.72) is generally true and can be evaluated for any of the cosmological models that we have described. We can obtain the essential features of light propagation by particularising to the Einstein–de Sitter model. We put $k = 0$ and $R = R_0(t/t_0)^{2/3}$ and integrate. This gives

$$r = \frac{ct_0^{2/3}}{R_0} \int_t^{t_0} \frac{dt}{t^{2/3}} = \frac{3ct_0^{2/3}}{R_0}(t_0^{1/3} - t^{1/3}). \tag{5.73}$$

Therefore the proper distance (the metrical distance, see (5.23)), of the photon from us at time t is given by

$$rR = 3ct_0^{1/3}t^{2/3} - 3ct, \tag{5.74}$$

where $R/R_0 = (t/t_0)^{2/3}$ has been used. Equation (5.74) is the world line of a light-ray. We have used it to plot our past light cone on the spacetime diagram shown in figure 5.6, where the time axis is cosmic time and the space axis is proper distance. Also plotted on the diagram is the world line of a particular galaxy g_1. Light from the galaxy reaching us at the present time is emitted at the point of intersection of our past light cone with the world line of the galaxy. If

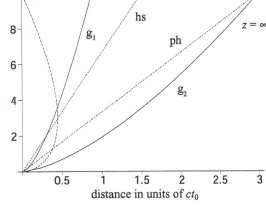

cosmic time in units of 10^9 years

Figure 5.6. Spacetime diagram showing our past light cone (dashed curve), Hubble sphere (broken, labelled hs) and particle horizon (broken, labelled ph). Our world line is the time axis. Also shown is the world line of a galaxy, or fundamental observer, observed now with redshift $z = 1$ (full line, labelled g_1). The world line labelled g_2 is that of a galaxy just entering our particle horizon at the present time.

this occurs at time t_e its proper distance from us as it crosses our past light cone is, from (5.74), given by

$$l_e = r_e R_e = 3ct_0^{1/3} t_e^{2/3} - 3ct_e, \qquad (5.75)$$

where the subscript e denotes the emission event. Alternatively, from (5.43), $t_e = t_0(1 + z)^{-3/2}$ so the proper distance of the galaxy at t_e as a function of its redshift is

$$l_e = r_e R_e = 3ct_0(1 + z)^{-1}(1 - (1 + z)^{-1/2}). \qquad (5.76)$$

The distance l_e goes through a maximum at $z = 5/4$ (problem 35). Note that a galaxy world line does not actually go back to the origin as galaxies were not formed until later.

5.18 The Hubble sphere

An interesting feature of the light cone in figure 5.6 is that it is not cone shaped but onion shaped. This tells us that a light wavefront (or photon) emitted in our direction before a time $t = 8t_0/27$ in an Einstein–de Sitter model (corresponding to a redshift greater than 5/4) initially recedes before it starts to approach us. The most extreme case is that of a photon emitted towards us just after $t = 0$.

The point of emission is almost on top of us, yet the light takes the present age of the Universe to reach us. This behaviour is caused by the recession from us of the space through which the light is propagating and is a consequence of the velocity–distance law (2.5).

This law gives the velocity, v_e, of a comoving galaxy at distance l_e:

$$v_e = H(t_e)l_e. \tag{5.77}$$

Thus substituting for l_e from (5.76) gives

$$v_e = 2c((1+z)^{1/2} - 1), \tag{5.78}$$

where $H(t_e) = H_0(1+z)^{3/2} = (2/3t_0)(1+z)^{3/2}$ has been used. Equation (5.78) tells us that when $z > 5/4$, then $v > c$. So the light is propagating in a region of space which is receding from us at a rate which is greater than the speed of light, thus, initially, the light is swept away from us until the expansion rate has slowed enough for the light to make ground in our direction (problem 36). This apparently odd behaviour of light is due to the very rapid expansion of space at early times. In the limit as $t \to 0$ the redshift goes to infinity and $v \to \infty$. As Eddington put it, the light is like a runner on an expanding track with the distance between the runner and the finishing line increasing faster than the runner can run.

The same conclusion can be drawn for any decelerating Universe (Harrison 1991). Galaxies at time t lying on the spherical surface of radius

$$l_H = \frac{c}{H} \tag{5.79}$$

are receding from us at exactly the speed of light. This surface, called the *Hubble sphere*, moves with velocity

$$\frac{dl_H}{dt} = c(1+q) \tag{5.80}$$

obtained by differentiating (5.79) with respect to time and using the definition of the deceleration parameter $q = -\ddot{R}R/\dot{R}^2$. For a decelerating Universe, where $q > 0$, the Hubble sphere overtakes galaxies at a relative velocity of cq. So a galaxy that is initially outside the Hubble sphere, and is receding at a speed $v > c$, will eventually be passed by the Hubble sphere and be receding at speed less than c. Similarly light propagating in our direction but losing ground will eventually be passed by the Hubble sphere and will start to approach us. But for what happens during an inflationary phase or in the steady-state Universe where $\rho = $ constant, see problem 40.

Superluminal velocities do not contradict the special theory of relativity and its stricture that c is an upper limit to all speeds, as this theory applies in the large only if there exists a global inertial frame of reference. In an expanding space we can set up a *local* inertial frame; so special relativity applies locally and the velocity of light relative to a local observer is always c. However, we cannot set up a global inertial frame to include both our own galaxy and a distant galaxy, as the spacetime is curved on this scale: so special relativity does not apply to such a situation. The bending of light in spacetime is equivalent to a tipping over of the local light cones and it is this that ensures compatibility between special and general relativity (figure 5.1).

5.19   The particle horizon

We now consider how the present distance, l_0, between a galaxy having radial coordinate r depends on z. From equation (5.73),

$$l_0 = r R_0 = 3ct_0^{2/3}(t_0^{1/3} - t^{1/3}) = 3ct_0(1 - (1+z)^{-1/2}), \qquad (5.81)$$

where (5.43) has been used. Thus the current distance from us of a galaxy with redshift z is a monotonically increasing function of z. The distinction between the distance from us at the time of emission and the distance from us now is made clear by figure 5.6. Galaxies on our past light cone sent out the light that reaches us now, but their present distance is given by the intersection of their world lines with the surface $t = t_0$. Of course, we see the galaxies as they were at the time of emission, not as they are now.

Since (5.81) shows distance to be an increasing function of redshift, it follows that the greatest distance from which we could, in principle, have received a signal is the present distance of a source having the largest possible redshift. We call this distance our *particle horizon* or simply our horizon if the context is clear. Letting $z \to \infty$ in (5.81) we see that in an Einstein–de Sitter Universe the distance of the particle horizon at time t is

$$D_h = 3ct. \qquad (5.82)$$

So at the present time $D_h(t_0) = 3ct_0 = 2c/H_0 = 6000h^{-1}$ Mpc. With $h = 0.65$ this gives $D_h(t_0) = 9200$ Mpc. The line $D_h = 3ct$ in figure 5.6 illustrates the growth of our horizon as a function of cosmic time. Any particle which crosses our past light cone at some time lies within our current horizon. It follows that any object beyond our particle horizon cannot have influenced us. From the spatial homogeneity of the Robertson–Walker metric nor can we have influenced it. We say that we are out of causal contact with the object. Particle horizons arise because the Universe has a finite age and the speed of light is finite. In a spatially flat static Newtonian Universe of age t we would be able to see to a maximum distance ct. In the expanding general relativistic models we can see further because, in a sense, the expanding space helps signals on their way.

We can obtain the horizon distance at time t for any FLRW model from equation (5.72) by setting the lower limit of the integral on the right-hand side equal to zero. This gives

$$\int_r^0 \frac{dr}{(1 - kr^2)^{1/2}} = -c \int_0^t \frac{dt'}{R(t')}. \qquad (5.83)$$

Multiply both sides of (5.83) by $R(t)$; from the definition of proper distance we get

$$D_{\rm h} = cR(t) \int_0^t \frac{dt}{R}. \tag{5.84}$$

All the FLRW models that we have discussed possess a particle horizon, because the integral on the right of (5.84) converges. In principle, one can investigate FLRW models which do not have horizons by choosing $R(t)$ such that the integral diverges, for example, the steady-state model (problem 41), although such behaviour requires a physically unrealistic matter content. The horizon distance is proportional to the time t, so the early Universe must have been made up of small causally disconnected regions. We will see in chapter 6 that there is a problem in reconciling such a disjointed early Universe with the smooth microwave background radiation which we observe.

5.20   Alternative equations of state

Details of the evolution of model Universes depend on assumptions about the matter content through the choice of equation of state. We shall see the importance of this in the early Universe in chapter 8. Here we explore some more esoteric alternatives.

Suppose first that the matter content is a gas of particles of *rest mass* density $\rho_{\rm m}$ and ratio of principal specific heats γ. The expansion is adiabatic, since there is no heat input, so

$$p \propto \rho_{\rm m}^{\gamma}. \tag{5.85}$$

We have $\rho_{\rm m} \propto R^{-3}$, so we can rewrite the equation of local energy conservation (5.26) in terms of $\rho_{\rm m}$ instead of t:

$$\rho_{\rm m} \frac{d\rho}{d\rho_{\rm m}} - \rho = \frac{p}{c^2}.$$

Then (5.85) yields

$$p \frac{d\rho}{dp} - \frac{\rho}{\gamma} = \frac{p}{\gamma c^2}$$

as a differential equation for $\rho(p)$. The constant of integration is determined by the condition $\rho \to \rho_{\rm m}$ when $p \to 0$. We obtain

$$\rho = \rho_{\rm m} + \frac{p}{(\gamma - 1)c^2},$$

with $\rho_{\rm m}$ given by (5.85). Until recently discussion was usually restricted either to the case of dust, $p = 0$, or to a highly relativistic fluid for which the internal energy dominates the rest mass, $\rho_{\rm m} \ll p/c^2$. An equation of state of the form

$$p = wc^2\rho$$

with $w = \gamma - 1$ covers all cases. Physical realism requires $\rho > 0$, which is known as the *weak energy condition*.

The speed of sound in a fluid is given by $v_s = |(dp/d\rho)|^{1/2}$. For $p > 0$ the condition that v_s lie between zero and the speed of light restricts w to the range

$$1 \geq w \geq 0.$$

The range of possibilities can be extended to include vacuum energy for which $p = -\rho c^2$. This gives

$$1 \geq w \geq -1. \tag{5.86}$$

Matter satisfying (5.86) is said to satisfy the *dominant energy condition* (because the density dominates the pressure, $\rho \geq |p|$). All known matter satisfies the dominant energy condition. For example, $w = 1/3$, $w = 0$, and $w = -1$ give the equations of state of radiation, dust and vacuum energy respectively. The case $w = 1$, for which $v_s = c$, is referred to as stiff matter (see section 8.4.2).

Measurements of supernovae at high redshift present strong evidence that the expansion of the Universe is speeding up (section 6.4.3). The deceleration parameter for a Universe containing matter and an unknown component of mass-energy, now called dark energy with equation of state $p = w\rho_d c^2$ is

$$q = \frac{\Omega_m}{2} + \frac{\Omega_d}{2}(1 + 3w). \tag{5.87}$$

Evidently a negative q requires a component of mass energy with negative pressure and $w < -1/3$. As the Universe evolves Ω_d falls less rapidly than Ω_m (see problem 39). So eventually the second term on the right in equation (5.87) will dominate the first and the deceleration will be negative. The obvious candidate for this dark energy is the cosmological constant and at the time of writing the observations are consistent with this form of dark energy (section 6.4.3). However, dark energy may arise from other sources of negative pressure. A favourite alternative candidate at present is an evolving scalar field with $-1 < w < -1/3$. This type of dark energy is referred to as quintessence. In principle, it should be possible to distinguish between a cosmological constant and quintessence (Perlmutter *et al* 1999).

5.21 Problems

Problem 22. *Show that the metric*

$$ds^2 = dx^2 + dy^2 \tag{5.88}$$

can be transformed into the form

$$ds^2 = dr^2 + r^2 \, d\theta^2 \tag{5.89}$$

by a suitable choice of coordinates. (Note that we do not have to specify that the coordinates in (5.88) are Cartesian coordinates. This information is contained in the form of the metric itself; similar remarks apply to the polar coordinates of expression (5.89).)

Problem 23. *The reader may be tempted to think of the FLRW models as comprising the ordinary spacetime of special relativity in which the fundamental observers are moving apart. Starting from the metric of the Milne model*

$$d\tau^2 = dt^2 - t^2[d\chi^2 + \sinh^2\chi\,(d\theta^2 + \sin^2\theta\,d\phi^2)],$$

make a transformation of coordinates

$$T = t\cosh\chi \qquad R = ct\sinh\chi$$

and hence show that the Milne model is equivalent to a part of flat spacetime (having $T > 0$, $R < T$). Show that the Milne model contains no matter (i.e. that $R(t) = ct$, $k = -1$, requires $\rho = 0$). In fact this, and the trivial case $R(t) = $ constant, are the only models which are equivalent to flat spacetime.

Problem 24. *The Lemaître redshift rule states that*

$$1 + z = \frac{R_0}{R_e} = \frac{R(t_0)}{R(t_0 - \Delta t)}.$$

By Taylor expansion of $R(t_0 - \Delta t)$ show that, for small Δt, $z = H_0\Delta t$. Hence, by using the velocity–distance law, show that $v \simeq cz$.

Problem 25. *Equation (5.76) gives the distance to a galaxy having redshift z at the time of emission and equation (5.81) gives the present distance to the same galaxy. Show that in the limit of small z both expressions reduce to Hubble's law $zc = H_0l$. In general this relation is only an approximation. However, in the steady-state theory it is exact (see problem 40).*

Problem 26. *Show that the static Einstein model Universe must have $k = +1$ and Λ positive, and that the density is given by*

$$\frac{k}{R^2} = \frac{4\pi G\rho}{c^2}.$$

Problem 27. *Show that the Friedmann equation and the acceleration equation are consistent with local energy conservation (i.e. derive equation (5.26) from (5.25) and (5.27)).*

Problem 28. *Typical velocities of galaxies (relative to the local standard of rest) are of the order of 300 km s^{-1}, so $v \ll c$. By treating the galaxies as an ideal gas* *show that* $3p/c^2 \ll \rho$. *From the acceleration equation (5.27) it is evident that the pressure of matter plays no role in the dynamics of the expansion.*

Problem 29. *Show that, in a Universe with* $\Lambda = 0$ *and k negative, curvature starts to dominate, that is the* $-kc^2$ *term becomes the dominant term on the right-hand side of the Friedmann equation, when*

$$R = \frac{R_0 \Omega_0}{(1 - \Omega_0)}.$$

Problem 30. *Carry out the demonstration that as* $\Omega_0 \to 1$ *equation (5.56) becomes* $t_0 \sim 2/(3H_0\Omega_0^{1/2})$. *Hint: For* Ω_0 *just greater than 1,* $(1 - 1/\Omega_0)^{1/2}$ *is small so we can use the expansion* $\sin^{-1} x \sim x + x^3/6$.

Problem 31. *It is useful to have the relation between the age t of a model and the corresponding redshift z. Show that for* $k = +1$ *this is*

$$t = \frac{\Omega_0}{H_0(\Omega_0 - 1)^{3/2}} \left\{ \sin^{-1} \left[\frac{(\Omega_0 - 1)}{(1 + z)\Omega_0} \right]^{1/2} \right.$$
$$\left. - \left[\frac{(\Omega_0 - 1)}{(1 + z)\Omega_0} \right]^{1/2} \left[1 - \frac{(\Omega_0 - 1)}{(1 + z)\Omega_0} \right]^{1/2} \right\}.$$

Hint: Start from equation (5.54). Show that for large z this is

$$t \sim \frac{2}{3H_0(1 + z)^{3/2}\Omega_0^{1/2}}.$$

In fact this result holds for all Ω_0 *(hence for all k).*

Problem 32. *Derive equation (5.69) by integrating the Friedmann equation with* $\rho = \rho_m + \rho_r$ *and with the curvature and cosmological terms set to zero. Hint: Note that* $\rho_R/\rho_M = (R_{eq}/R)(R/R_0)$.

Problem 33. *Using equations (5.54) and (5.32) show that the total lifetime of a* $k = +1$ *model is given by*

$$t_{\text{lifetime}} = \frac{\pi \Omega_0}{H_0(\Omega_0 - 1)^{3/2}}.$$

Problem 34. *For a critical density* $k = 0$ *model with* $\Lambda \neq 0$ *obtain the expression for the age (section 5.15.3)*

$$t_0 = \left(\frac{2}{3H_0} \right) \frac{\tanh^{-1} \Omega_\Lambda^{1/2}}{\Omega_\Lambda^{1/2}}.$$

Problem 35. *Show that, in an Einstein–de Sitter model, the proper distance l_e at the time of emission is a maximum for a galaxy with redshift $z = 5/4$.*

Problem 36. *Use the Robertson–Walker metric to show that for a photon moving radially towards the origin of coordinates*

$$\frac{dl}{dt} = Hl - c,$$

where l is the proper distance of the photon from the origin at time t. Hint: $l = R(t)f(r)$ where $f(r) = \int dr/(1 - kr^2)^{1/2}$.

Problem 37. *(a) Show that the rate of change at the present time of the redshift of a galaxy is given by*

$$\frac{dz}{dt_0} = H_0(1 + z) - H(z).$$

Hint: Differentiate $1 + z = R_0/R$ with respect to t_0 and note that $dt_0/dt_e = (1+z)$ from cosmological time dilation.
    (b) Show that dz/dt_0 is negative in an Einstein–de Sitter Universe.
    (c) Show that in a Universe having constant density $dz/dt_0 = H_0 z$.

Problem 38. *Show that $\Omega H^2 R^3 = $ constant in a matter-dominated Universe.*

Problem 39. *For a model with general equation of state $p = w\rho c^2$ use the equation of local energy conservation to show that $\rho \propto R^{-3(1+w)}$.*

Problem 40. *In the steady-state Universe $\rho = $ constant. (a) Show that the rate of creation of new matter needed to keep ρ constant is $3\rho H$. (b) Assuming that H is constant in the steady-state Universe use the Friedmann equation to show that the spatial geometry is Euclidean. (c) Show that the scale factor $R(t) \propto \exp(Ht)$. Hence deduce from equation (5.80) that the Hubble sphere is stationary. (d) By considering the path of a light-ray from a galaxy to us show that*

$$zc = Hd_0$$

exactly, where d_0 is the present distance of the source. (e) Show that our past light cone asymptotically approaches the Hubble sphere. Hint: Compare the treatment of the Einstein–de Sitter model in the text.

Problem 41. *Show that the steady-state Universe has an event horizon but no particle horizon. Hint: Use the previous question.*

Chapter 6

Models

6.1 The classical tests

We have seen in the last chapter how the general theory of relativity, together with the cosmological principle, leads to a set of mathematical models for the Universe. These models describe the Universe in terms of a spacetime metric from which the motion of bodies and the paths of light-rays can be determined. The models are specified by four quantities: the density parameter Ω_M of the matter (including both baryonic matter and dark matter), the density Ω_R of radiation, the cosmological constant Λ, or the equivalent vacuum density parameter Ω_Λ, and the Hubble constant H_0. A major goal of cosmology is to decide which of these models, if any, describes the actual Universe. In order to answer this question Kohler, Tolman, Hubble and others in the 1930s devised what, on account of their venerability, are now known as the classical cosmological tests. Since then additional tests, sometimes referred to as the neo-classical tests, have been added to the original ones.

Many quantities entering into the models, such as proper distances of galaxies and their velocities of recession, cannot be measured directly. The quantities that we can measure are the redshifts, magnitudes, and angular sizes of galaxies and their surface brightnesses. We can also count galaxies. In order to test the models against observation, relationships between these measurable quantities must be derived from the metrics. These relationships are the basis of the cosmological tests that form the subject of this chapter. The tests that we shall consider are:

(1) the angular diameter of a standard rod as a function of redshift;
(2) the Hubble plot in which the apparent magnitude of a standard candle as a function of redshift is determined;
(3) the quasar lensing test;
(4) the number–redshift test;
(5) the timescale test.

In principle, any one of the first four of these tests allows us to determine Ω_M and Ω_Λ. The tests also provide an independent check on the value of Ω_M which was obtained in chapter 3 from the typical mass-to-light ratio of clusters. If one of the relativistic models does describe the actual Universe then the whole battery of tests must give concordant results. On the other hand, if the different tests give conflicting results we shall have to conclude that none of the relativistic models provides an adequate description of the Universe. In that case the theoreticians will have to go back to the drawing board.

It is important to understand that general relativity provides a complete picture of the behaviour of physical systems in gravitational fields. The outcomes of the classical tests are therefore completely determined once we assume that our physical laws are valid throughout the Universe (the Copernican principle of chapter 1). This is why serious workers in the field start from the expanding Universe models: if they disagree with observation one learns something of importance. It is possible to account for cosmological observations one by one in various ingenious ways that require neither relativity nor expanding space. These *ad hoc* models are not interesting: if one of them does not work one just moves on to the next alternative remedy.

All of the relativistic models describe an expanding Universe. As we have seen in chapter 4, a consequence of the expansion is that light becomes redshifted as it travels towards us. Not only is the light redshifted but all physical processes taking place at the source are observed to be slowed down by a factor of $(1 + z)$. This effect is known as cosmological time dilation. We shall see in section 6.10 that cosmological time dilation has been observed. This rules out *ad hoc* explanations of the redshift, such as the 'tired light' hypothesis which accounts only for the redshift of photons, and provides support for the expanding Universe picture.

6.2   The Mattig relation

Let us suppose that an observer is situated at the origin of coordinates in a Robertson–Walker Universe with metric

$$ds^2 = c^2 \, dt^2 - R^2(t) \left\{ \frac{dr^2}{(1 - kr^2)} + r^2 (d\theta^2 + \sin^2 \theta \, d\phi^2) \right\}. \qquad (6.1)$$

To apply the cosmological tests we shall need a relation between the redshift of a galaxy and its radial coordinate.

Consider an incoming radial ray of light from some emission event. Along the ray we shall have $d\theta = d\phi = 0$, since the ray is radial, and $ds = 0$ since it is a light-ray. Integrating along the ingoing ray from the emission event $r = r_e$ at time t to the reception event $r = 0$ at time t_0 gives (see section 5.17)

$$-\int_{r_e}^{0} \frac{dr}{(1 - kr^2)^{1/2}} = c \int_{t}^{t_0} \frac{dt}{R}. \qquad (6.2)$$

Evaluating the integral on the left-hand side gives

$$\chi = \int_0^{r_e} \frac{dr}{(1 - kr^2)^{1/2}} = k^{-1/2} \sin^{-1}(k^{1/2}r_e). \tag{6.3}$$

The result is valid in all three cases, $k = 0, \pm 1$. To obtain an appropriate form for $k = 0$ we use

$$\lim_{k \to 0} [k^{-1/2} \sin^{-1}(k^{1/2}r_e)] = r_e$$

and for $k = -1$ we make the replacement $i^{-1} \sin^{-1} i r_e = \sinh^{-1} r_e$.

We now have to replace the integral over time on the right-hand side of (6.2) by an integral over redshift, z. We start from the fundamental relation between time and redshift,

$$1 + z = \frac{R_0}{R}.$$

Differentiating gives

$$\frac{dz}{dt} = -\frac{R_0}{R^2} \frac{dR}{dt} = -\frac{R_0 H}{R}$$

so

$$\frac{dt}{R} = -\frac{dz}{R_0 H(z)}. \tag{6.4}$$

6.2.1   The case $p = 0$, $\Lambda = 0$

Before we consider the general situation we take as an example the case of a Universe with $p = 0$ and $\Lambda = 0$. Here $H(z)$ is given by equation (5.40):

$$H(z) = H_0(1 + z)(1 + \Omega_0 z)^{1/2},$$

where Ω_0 is the current value of the total matter density. The right-hand side of (6.2) becomes

$$\chi = c \int_t^{t_0} \frac{dt}{R} = -\frac{c}{R_0} \int_z^0 \frac{dz'}{H(z')} = \frac{c}{R_0 H_0} \int_0^z \frac{dz'}{(1 + z')(1 + \Omega_0 z')^{1/2}}. \tag{6.5}$$

The integration can be carried out using the substitution $y = (1 + z)^{-1}$. After a lot of algebra we get

$$\chi = k^{-1/2} \sin^{-1} \left\{ \frac{2k^{1/2}c[z\Omega_0 + (\Omega_0 - 2)((\Omega_0 z + 1)^{1/2} - 1)]}{R_0 H_0 \Omega_0^2 (1 + z)} \right\}. \tag{6.6}$$

Combining (6.6) with (6.3) gives our result:

$$R_0 r_e = 2c(H_0 \Omega_0^2 (1 + z))^{-1} [z\Omega_0 + (\Omega_0 - 2)((\Omega_0 z + 1)^{1/2} - 1)], \tag{6.7}$$

which is known as the Mattig relation.

6.2.2   The general case $p = 0$, $\Lambda \neq 0$

In general from equations (6.2) to (6.4), we have

$$R_0 r_e = R_0 k^{-1/2} \sin\left[(k^{1/2}c/R_0)\int_0^z \frac{\mathrm{d}z'}{H(z')}\right]. \qquad (6.8)$$

If $\Lambda \neq 0$ we have

$$H(z) = H_0[(1+z)^2(1+z\Omega_M) - z(2+z)\Omega_\Lambda]^{1/2}, \qquad (6.9)$$

where Ω_M and Ω_Λ are, as usual, the density parameters at the present time arising from the matter and the cosmological constant respectively. For the derivation of (6.9) see problem 43. For the case $\Lambda = 0$ we have shown that equation (6.8) integrates to give the Mattig relation (6.7). For non-zero Λ, however, equation (6.8) must be evaluated numerically. Note that the more general case $p \neq 0$, $\Lambda \neq 0$ is of academic interest only, since in the early Universe when the pressure is significant the contribution from the cosmological constant, $\Omega_\Lambda = \Lambda/3H^2$, is small.

6.3   The angular diameter–redshift test

6.3.1   Theory

Consider again an observer situated at the origin of coordinates in a Robertson–Walker Universe with metric (6.1). Let our observer measure the small angular size, $\Delta\phi$, of the major axis of a galaxy which, for the sake of argument, lies in the plane $\theta = \pi/2$. Let t_e be the time of the emission of the light from the galaxy towards us that we receive now. Then, from the metric (6.1) with $\Delta t = \Delta\theta = \Delta r = 0$, the proper distance between the points $(r, \pi/2, \phi)$ and $(r, \pi/2, \phi + \Delta\phi)$ is $r R(t_e)\Delta\phi$. So, if D is the diameter of the galaxy, we may write

$$D = r_e R_e \Delta\phi, \qquad (6.10)$$

for sufficiently small $\Delta\phi$. Figure 6.1(*a*) shows that the angle $\Delta\phi$ between the two rays at emission is equal to the angle between the two rays at reception. Figure 6.1(*b*) shows the light paths on a space time diagram.

From (6.10)

$$\Delta\phi = \frac{D}{r_e R_e} = \frac{D(1+z)}{r_e R_0}, \qquad (6.11)$$

where the last step follows because $R_0/R_e = 1 + z$. Consider the case $p = 0$, $\Lambda = 0$. Using the Mattig relation (6.7) for $r_e R_0$ as a function of the redshift z of the galaxy gives

$$\Delta\phi = \frac{D(1+z)^2 H_0 \Omega_0^2}{2c[z\Omega_0 + (\Omega_0 - 2)((\Omega_0 z + 1)^{1/2} - 1)]}. \qquad (6.12)$$

(a)

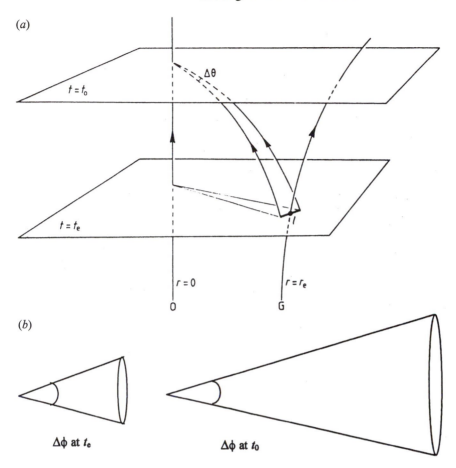

(b)

$\Delta\phi$ at t_e

$\Delta\phi$ at t_0

Figure 6.1. (*a*) Spacetime diagram showing the rays from the edges of a galaxy G, emitted at time t_e and observed by O at time t_0. (*b*) Uniform expansion does not alter angles.

This expression tells us that for small z, $\Delta\phi \propto 1/z$, the behaviour to be expected in a static Universe. Thereafter $\Delta\phi$ reaches a minimum for some finite value of z, after which it increases monotonically (see figure 6.2). In the Einstein–de Sitter case this minimum occurs at $z = 5/4$ (see problem 44). The occurrence of a minimum angular size is a consequence of the expansion of space which features in all the FLRW models. We saw in section 5.17 that, beyond a certain value of z, the distance l_e, in equation (5.76), of a galaxy at the time that light left it, ceases to grow and gets smaller with increasing z. This means that as we look to higher redshifts our standard rod gets closer to us at emission and so subtends larger angles at our location. Thus the detection of a minimum in the

$\Delta\phi$–z relationship would be strong evidence that the redshift of galaxies really is due to the expansion of the Universe.

We can see that (6.12) appears to allow $\Delta\phi$ to exceed 2π for some z; this would mean that the size of our rod exceeds the size of the space available to it at this epoch. However, once $\Delta\phi$ becomes large, as $z \to \infty$, the formula (6.12) is no longer appropriate, since it was derived on the basis of a small $\Delta\phi$. A small $\Delta\phi$ is required so that we do not need to distinguish between an arc, $r = $ constant, and a chord in equation (6.10).

6.3.2   Observations

In practice we cannot accurately measure the geometric diameter of a galaxy at a large redshift by optical means. This is because the visibility of the faint outer parts of a galaxy against the background depends on redshift. An apparently more suitable 'standard rod', suggested by Hoyle in 1958, is the angular separation of the two lobes of a radio galaxy. When it became possible to carry out such measurements, starting in the 1970s, it was found that $\Delta\phi \propto 1/z$ up to large redshifts in disagreement with expression (6.12). However, these radio sources are much larger than their host galaxies, so the separation between their lobes, which are created by jets of material shot out of the radio galaxy, must depend on conditions in the intergalactic medium at the time of their formation. It therefore seems likely that the linear size of these radio sources was smaller in the past, as the intergalactic medium was denser in the past, and they are therefore not good 'standard rods'. Consequently the observations have to be corrected for this evolutionary effect before they can be used to test cosmological models, something which, at present, cannot be done.

A more recent measurement of the angular size–redshift relation has used the jets in the nuclei of quasars as the 'standard rod'. These jets are more compact and lie within the host galaxy, so their lengths are less likely to change with cosmic epoch. They can be mapped by using very long baseline radio interferometry which gives milli-arcsecond resolution. The results of this study (Kellermann 1993) are shown in figure 6.2. This is the first time that a minimum in the $\Delta\phi$–z relationship has been detected. However, these results have been criticized; see for example Dabrowski *et al* (1997). It is not yet possible to use these results to put significant constraints on cosmological models.

Figure 6.2 shows plots of angular size as a function of redshift for different values of $q_0 = \Omega_0/2$. There is also some more recent work on this test (e.g. Guerra and Daly 1998).

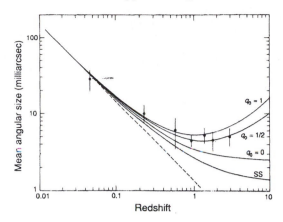

Figure 6.2. Mean angular size plotted against redshift for 82 compact radio sources (points with error bars) and for various FLRW models and the steady-state model (SS). Individual points are based on a number of sources (adapted from Kellermann 1993).

6.4   The apparent magnitude–redshift test

6.4.1   Theory

The FLRW models give us a relationship between the apparent magnitude, m, and the redshift, z, of a 'standard candle' source. By 'standard candle' we mean a class of objects which all have the same absolute magnitude M. By definition, the absolute magnitude of a source is related to its apparent magnitude and its distance, l, by

$$m - M = 2.5 \log(l/10 \text{ pc})^2. \tag{6.13}$$

For small redshift z the dependence of the distance l on z, is given by Hubble's law (chapter 2)

$$cz = H_0 l.$$

Substituting for l from Hubble's law into (6.13) gives the following linear relationship:

$$\log cz = 0.2m + \text{constant}. \tag{6.14}$$

Note that the magnitude of the slope in (6.14) depends on the power to which the distance l in Hubble's law is raised. If this power were to be n rather than unity, equation (6.14) would have a slope of $0.2n$. The measured slope of the so called Hubble plot gives $n = 1.005 \pm 0.018$ (Jones and Fry 1998). This agreement supports the assumption of uniform expansion (see section 2.2) and rules out proposals such as a quadratic Hubble law.

We now consider the form of the Hubble plot when it is extended to higher redshifts. As we shall show the plot ceases to be linear and its exact form depends on both Ω_0 and Ω_Λ, so it provides us with a means of determining both these quantities.

Consider a standard candle source, which we place at the origin of the Robertson–Walker coordinate system, and suppose that we are located at radial coordinate r. Note that from the assumption of homogeneity only convenience determines where we place the origin of our coordinates. Let the total energy per second emitted by the source, its bolometric luminosity, be denoted L_{bol}. Then the total flux of radiation, that is the energy per unit area, crossing the spherical surface at radial coordinate r is given by

$$F_{bol} = \frac{L_{bol}}{4\pi R_0^2 r^2 (1+z)^2}. \tag{6.15}$$

The factor $4\pi R_0^2 r^2$ is the proper area of the surface through which the radiation passes, as follows from the metric (6.1). The first factor of $(1+z)$ arises because each photon suffers an energy loss due to the redshift of the radiation; the second factor of $(1+z)$ is due to the reduction in the arrival rate of photons. (In section 5.10 we showed that all physical processes in a distant galaxy are seen slowed down.) Now the energy flux from a source depends on the frequency of individual photons and on their rate of arrival. Both of these frequencies are slowed down so there are two factors of $(1+z)$.

Now the distance modulus expressed in terms of bolometric flux F is

$$m_{bol} - M_{bol} = 2.5\log(F_1/F_{bol}), \tag{6.16}$$

where $F_1 = L_{bol}/[4\pi(10 \text{ pc})^2]$. Combining (6.15) with (6.16) gives

$$m_{bol} - M_{bol} = 2.5\log[(R_0 r)^2 (1+z)^2/(10 \text{ pc})^2]. \tag{6.17}$$

Expressing distances in Mpc this may be written as

$$m_{bol} - M_{bol} = 5\log[(R_0 r)(1+z)] + 25, \tag{6.18}$$

where $10 \text{ pc} = 10^{-5}$ Mpc has been used.

The next step is to express $R_0 r$ as a function of the redshift z of the source. This is given by the Mattig relation where now, for non-zero Ω_Λ, equation (6.8) must be evaluated numerically.

Figure 6.3 shows a plot of m against $\log cz$ for a number of different choices of Ω_M and Ω_Λ.

6.4.2   The K-correction

Note that equation (6.18) expresses bolometric magnitude as a function of redshift. In practice detectors operate over a restricted range of frequencies so they only measure a fraction of the total luminosity; moreover, this fraction will vary with the redshift. There are two effects at work. First, the range of

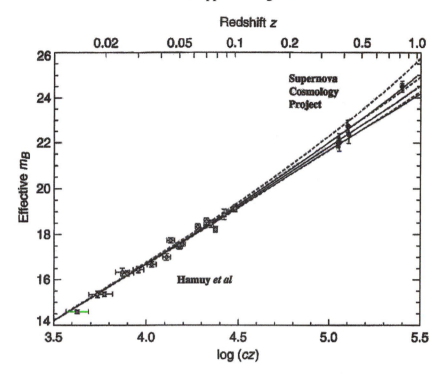

Figure 6.3. Hubble plot of effective magnitude against $\log cz$. The full curves are computed from models for various values of Ω_M and Ω_Λ with, from the top down, $(\Omega_M, \Omega_\Lambda) = (0,0)$, $(1,0)$, $(2,0)$ and similarly for the dotted curves $(\Omega_M, \Omega_\Lambda) = (0,1)$, $(0.5, 0.5)$, $(1,0)$, $(1.5, -0.5)$. The experimental points are from type 1a supernovae (Perlmutter *et al* 1998).

frequencies in the rest frame of the source, registered by the detector, is increased by a factor of $(1 + z)$; this makes the source appear brighter than it should by this factor. The second effect arises because photons observed at frequency ν_0 were emitted at frequency $(1 + z)\nu_0$. So the luminosity of the source measured in a band centred on frequency ν_0 is $L((1 + z)\nu_0)$ rather than $L(\nu_0)$. Only if the source has a completely flat spectrum with intensity independent of frequency would this make no difference.

To take account of these two effects, let the energy emitted per unit time in the frequency range ν_e to $\nu_e + d\nu_e$ at the source be $L(\nu_e)\,d\nu_e$. Then the flux at the observer corresponding to this luminosity is

$$F(\nu_0)\,d\nu_0 = \frac{L(\nu_e)\,d\nu_e}{4\pi R_0^2 r^2 (1 + z)^2},$$

as in (6.15). But $\nu_e = (1+z)\nu_0$ so

$$F(\nu_0)\, d\nu_0 = \frac{L((1+z)\nu_0)(1+z)\, d\nu_0}{4\pi R_0^2 r^2 (1+z)^2}.$$

Integrate this over the bandwidth of the detector (assuming for simplicity that the sensitivity is constant over the band) to obtain the flux at the detector:

$$F_0 = \int_B F(\nu_0)\, d\nu_0 = \int_B \frac{L((1+z)\nu_0)(1+z)\, d\nu_0}{4\pi R_0^2 r^2 (1+z)^2} \tag{6.19}$$

$$= \frac{\int_B L(\nu_0)\, d\nu_0}{4\pi R_0^2 r^2 (1+z)^2} \times \frac{\int_B L[(1+z)\nu_0]\, d\nu_0 (1+z)}{\int_B L(\nu_0)\, d\nu_0}.$$

The flux in the band at 10 pc would be

$$F_1 = \frac{\int_B L(\nu_0)\, d\nu_0}{4\pi (10\,\mathrm{pc})^2}.$$

Thus the theoretical $m(z)$ relation becomes

$$m - M = 5\log[(R_0 r)(1+z)] + 25 - K(z), \tag{6.20}$$

where the corrections for finite bandwidth have been lumped together in $K(z)$, called the K-correction, and

$$K(z) = 2.5\log[f(z)(1+z)]$$

with

$$f(z) = \frac{\int_B L[(1+z)\nu_0]\, d\nu_0}{\int_B L(\nu_0)\, d\nu_0}$$

(see problem 46). The parameters Ω_M and Ω_Λ can be determined, in principle, by fitting this expression to an m–z plot obtained from observation.

6.4.3 Magnitude versus redshift: observations

To carry out the test requires a standard candle source bright enough for its redshift to be measured out to a large value, which means a redshift of the order of unity or greater. Arbitrarily chosen galaxies have widely different luminosities so are not suitable as standard candles. The brightest galaxies in rich clusters, however, show a much smaller variation in intrinsic luminosity and have until recently been the preferred choice for the test. They possess the drawback that at a redshift $z \sim 1$ they are seen when the Universe was about one-third of its present age. We therefore have to take account of the expected evolution in the luminosity of the galaxies. Unfortunately we do not have an accurate model of galaxy evolution (Sandage 1988) so it has not been possible to untangle the effects of spacetime geometry from those of the evolution of the sources.

The breakthrough, that has occurred in the last few years has come from using supernovae of type Ia as the standard candles. There are two principal categories of supernovae (SNe): type I which lack hydrogen in their optical spectrum and type II which exhibit hydrogen lines. More recently in the 1980s it was recognized that type I SNe can be further subdivided into SNe Ia and SNe Ib and Ic. It is now known that type II SNe are produced by the core collapse of a massive star. It is believed that SNe Ib/Ic also arise from the core collapse of a massive star, but one which has lost its outer layers, hence the lack of hydrogen. On the other hand, supernovae of type Ia are believed to be the result of the thermonuclear explosion of a carbon–oxygen white dwarf star initiated when the star's mass is pushed over the Chandrasekhar limit, approximately $1.4M_\odot$, by accretion from a companion star (Filippenko 1997).

SNe Ia have the three important features required of a standard candle: they are very luminous ($M_B \sim -19.2$) so they can be seen out to large distances and, once so called 'peculiar' SNe Ia have been excluded, the dispersion of their absolute magnitudes about the mean is small. The range of absolute magnitudes is determined from a local sample of SNe Ia the distances to which are obtained using Cepheid variable stars in their host galaxies. Finally and, very importantly, it appears that the physical characteristics of low and high redshift SNe Ia are very similar, so there is no indication, at the time of writing, of any systematic evolutionary effects.

A Hubble plot of apparent magnitude against redshift for SNe Ia shows some scatter about a smooth curve. This scatter occurs because SNe Ia do not all have exactly the same absolute magnitudes. However, some of the scatter can be removed by using the light curves of the supernovae. The initial rate of decline of the optical light curve beyond its peak has been found to correlate strongly with its peak luminosity (Phillips 1993): the brighter the supernova at peak luminosity the slower the decay. Thus the light curve can be used to correct the peak apparent magnitudes m_B to give effective magnitudes which all correspond to a single absolute magnitude M, with an uncertainty of 17% in flux (Perlmutter *et al* 1999). Doing this reduces the scatter in the Hubble plot: see figure 6.3. These advances have made possible, for the first time, the determination of both Ω_M and Ω_Λ from the apparent magnitude–redshift test.

At the time of writing two research teams have published determinations of Ω_M and Ω_Λ based on the m–z plot using SNe Ia as the standard candles. They are the Supernova Cosmology Project (Perlmutter *et al* 1999) and the High-z Supernova Search (Riess *et al* 1998). Figure 6.4 from Perlmutter *et al* shows the confidence regions in an Ω_Λ–Ω_M plot. The first conclusion they draw is that an Einstein–de Sitter Universe is effectively ruled out as the point $\Omega_M = 1$, $\Omega_\Lambda = 0$ lies well below the 99% contour. Second, even an empty Universe with zero cosmological constant is ruled out at the 99.8% confidence level (see also figure 6.4). The data strongly favour $\Omega_\Lambda > 0$. If $\Omega_0 = 1$, as is indicated by observations of the microwave background (chapter 9), then the plot gives $\Omega_M \simeq 0.3$ and $\Omega_\Lambda \simeq 0.7$. With these values for Ω_M and Ω_Λ the deceleration

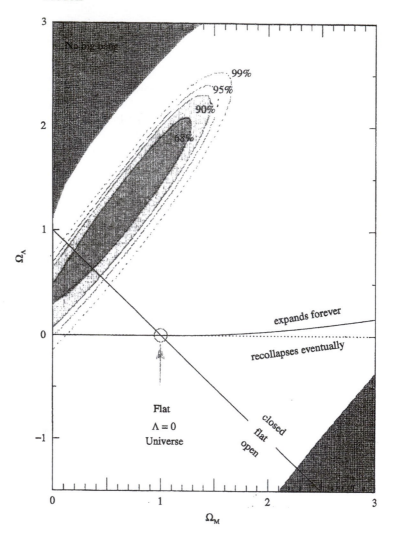

Figure 6.4. Allowed ranges of Ω_M and Ω_Λ showing the confidence regions obtained from observations of type 1a supernovae by the Supernova Cosmology Project (Goldhaber and Perlmutter 1998).

parameter is

$$q_0 = \tfrac{1}{2}\Omega_M - \Omega_\Lambda = -0.56 \tag{6.21}$$

(problem 47). This means that the rate of expansion is speeding up and, if correct, tells us that the Universe is destined to expand forever. The age of such a critical density Universe is given by equation (5.64). With $\Omega_\Lambda = 0.7$ this yields an age of 14.5 Gy.

Note from equation (6.9) that a positive Ω_Λ lowers the value of $H(z)$ which increases $r R_0$ and hence the magnitude m. Thus the effect of a cosmological constant is to make the SNe Ia at a given redshift fainter. In order to be confident that the observed faintness really is due to a cosmological constant it is important to eliminate other non-cosmological explanations. For example could the faintness be due to obscuration by dust in the host galaxy? The effect of dust would be wavelength dependent and would produce a reddening of the supernova's light. No such reddening, compared with local examples, has been observed. Another possibility mentioned previously is that the SNe Ia are systematically less luminous at early times as a result of differences in composition. No evidence for this has been found either.

The detection of a cosmological constant of magnitude $\Omega_\Lambda \sim 0.7$ is very interesting as it corresponds to a mass density which, at the present time, is nearly the same as that of the matter. Remember that $\rho_m \propto R^{-3}$ varies over many orders of magnitude with time and that $\rho_\Lambda = \Lambda/(8\pi G) = \text{constant}$. So this near equality of mass densities at the present appears to require an explanation; none is at present forthcoming. It is possible that the contribution to the total Ω which has been attributed to a cosmological constant could arise from some as yet unknown source of mass density (Perlmutter *et al* (1999) and see section 5.20).

Note that $\Omega_M \sim 0.3$ is in agreement with the value we quoted in chapter 4 obtained from the mass-to-luminosity ratios of clusters. But recall that that method does not detect any uniformly distributed mass density so does not measure the contribution of Ω_Λ. The m–z test is sensitive to the spatial geometry and to the expansion history so depends on both Ω_Λ, Ω_M and on any other sources of mass density. In chapter 9 we will see that the angular scale of the temperature fluctuations in the microwave sky is governed by the spatial curvature. The planned MAP and Planck satellites should provide stronger constraints on the values of $\Omega_0 = \Omega_\Lambda + \Omega_M$ and also of Ω_M. Current measurements from balloon-borne experiments give $\Omega_0 \approx 1$ (de Bernardis *et al* 2000).

6.5   The geometry of number counts: theory

In this section we investigate how the number of galaxies in a region of the sky depends on the redshift of the galaxies. Our starting point is given by the Robertson–Walker metric, equation (6.1). When we observe galaxies we see them as they were at the time t when their light set out to us: this is the *world picture*. So those galaxies within the solid angle $d\omega$ with radial coordinate in the range r to $r + dr$ occupy the following element of proper volume at time t:

$$dV = \frac{R(t)^3 r^2}{(1 - kr^2)^{1/2}} \, d\omega \, dr, \tag{6.22}$$

where $d\omega = \sin\theta \, d\theta \, d\phi$. We now write this volume element in terms of redshift z by expressing r as a function of z. Along an approaching radial null-ray we have

$$\frac{dr}{(1 - kr^2)^{1/2}} = -c\frac{dt}{R} = c\frac{dz}{R_0 H(z)}, \tag{6.23}$$

with $H(z)$ given by equation (5.40), where, for simplicity, we are confining ourselves to models with $\Lambda = 0$. Substituting into (6.22) from (6.23), substituting for the r^2 term from the Mattig relation (6.7), and integrating over a solid angle gives

$$\frac{dV}{dz} = 2\pi \left(\frac{2c}{H_0}\right)^3 \frac{(\Omega_0 z + (\Omega_0 - 2)\{-1 + (1 + \Omega_0 z)^{1/2}\})^2}{\Omega_0^4 (1 + z)^6 (1 + \Omega_0 z)^{1/2}}. \tag{6.24}$$

For small z the volume element increases in magnitude at the rate,

$$\frac{dV}{dz} = 4\pi \left(\frac{c}{H_0}\right)^3 z^2, \tag{6.25}$$

which is the rate of increase of volume with distance $d = cz/H_0$ expected in Euclidean space. Thereafter the volume element grows with increasing z at a diminishing rate until beyond $z \sim 1$ it starts shrinking in size (see problem 48). Part of the explanation for this behaviour is that, beyond a certain redshift, the distance between us and a galaxy actually decreases with z, as was explained in section 5.17.

We now consider the effect of this closing up of the volume element on the count rate. The number of counts per unit redshift is related to dV/dz as follows:

$$\frac{dN}{dz} = n\frac{dV}{dz}, \tag{6.26}$$

where n is the number density of galaxies at time t. Assuming that the number of galaxies is conserved, $nR^3 = n_0 R_0^3$ and so $n = n_0(1 + z)^3$, where n_0 is the number density of galaxies at the present time. Thus the number of galaxies ΔN in the redshift range Δz at redshift z is

$$\Delta N = \left(\frac{dN}{dz}\right)\Delta z = n_0(1 + z)^3 \left(\frac{dV}{dz}\right)\Delta z. \tag{6.27}$$

Despite the extra factor of $(1 + z)^3$ in the numerator, ΔN also falls at large z.

The effect of a non-zero cosmological constant is to increase the volume available at high redshift. In what follows for simplicity we confine the discussion to $\Lambda = 0$ models.

6.5.1   Number counts: observations

In principle we can count the total number of galaxies in a redshift interval Δz at z within the solid angle $\Delta\omega$ and then use equation (6.27) to obtain Ω_0. The practice, however, is not so straightforward. It is not possible to count all the galaxies in a redshift interval at an appreciable value of z: our count will miss very faint galaxies. However, we can count the number of galaxies brighter than some limiting flux $F_{\min}$ at the detector. The theoretical number of galaxies $n(> F_{\min})\Delta z$ between redshifts z and $z + \Delta z$ that exceed this limiting flux can, in principle, then be worked out from

$$n(> F_{\min})\Delta z = \int_{L_{\min}(z)}^{\infty} \Phi(L, z)\, dL\, \frac{dV}{dz} \Delta z, \qquad (6.28)$$

if we know the galaxy luminosity function $\Phi(L, z)$ as a function of redshift z. This function is defined such that $\Phi(L, z)dL\, dz$ is the number of galaxies per unit volume with luminosities in the range $(L, L+dL)$ in the redshift range $(z, z+dz)$. The lower limit to the integral in (6.28), $L_{\min}(z)$, is obtained as follows. The chosen minimum flux at the detector in the waveband observed, $F_{\min}$ gives the corresponding luminosity in this band from equation (6.19). If we assume that the number of galaxies is conserved, so $n = n_0(1 + z)^3$, and that the luminosity of a galaxy does not evolve with time, then

$$\Phi(L, z) = \Phi(L)(1 + z)^3$$

so

$$n(> F_{\min})\Delta z = \int_{L_{\min}(z)}^{\infty} \Phi(L)\, dL (1 + z)^3 \frac{dV}{dz} \Delta z.$$

The problem with implementing this test to obtain Ω_0 is the calculation of $n(> F_{\min})$. Loh and Spillar (1986) assumed the conservation of galaxies, and also that the galaxy luminosity function has the Schechter form (section 3.3) at all redshifts, and obtained $\Omega_0 = 1^{+0.7}_{-0.5}$. However, these assumptions are violated in simple models of galaxy evolution (Bahcall and Tremaine 1988) so the use of this test to obtain the cosmological parameters awaits a detailed model for galaxy evolution.

6.5.2 The galaxy number–magnitude test

This test was first used by Hubble in 1926. Two years previously he had established that galaxies were star systems lying beyond the Milky Way and he used the test to probe the distribution of the galaxies. In this test we count all sources that exceed a given flux irrespective of their redshift. Assume first, for simplicity, that all galaxies have the same absolute magnitude M_*. Suppose that, in the region surveyed, the distribution is uniform with a number density of galaxies n per unit volume. Then the number of galaxies counted that are brighter than apparent magnitude m equals the number within a sphere of radius r_{M_*} pc, say:

$$N(< m) = N(< r_{M_*}) = \tfrac{4}{3}\pi r_{M_*}^3 n.$$

The radius r_{M_*} is determined by the fact that a galaxy at this distance has the limiting apparent magnitude m. From the definition of absolute magnitude

$$m - M_* = 5 \log r_{M_*} - 5.$$

Eliminating r_{M_*} we obtain

$$\log N(< m) = 0.6m + \text{constant}. \tag{6.29}$$

To take account of the different absolute magnitudes of galaxies we replace the density n by the number of galaxies per unit volume in each magnitude range and integrate over magnitude

$$N(< m) = \int \Phi(L) \frac{dL}{dM} \left(\frac{4}{3} \pi r_M^3 \right) dM.$$

Eliminating r_M gives an expression of the form (6.29) with only a different constant. This expression was derived for a non-expanding Euclidean Universe so applies only to galaxies at low redshift. If the test is extended to fainter magnitudes we expect to see a departure from this simple law arising from the expansion of the Universe and from spatial curvature. So the slope $d \log N(< m)/dm$ will depart from the value of 0.6 and the plot will depend on the parameters Ω_M and Ω_Λ. In 1936 Hubble extended his observations to fainter magnitudes and claimed to have detected spatial curvature (Sandage 1988). His claim was premature and it is still not possible to use this test to determine the cosmological parameters.

The following factors will determine the make-up of the counts. From equation (6.19) the flux from a source of given luminosity decreases monotonically with z. So a high fraction of nearby galaxies will be counted. But the fraction of galaxies counted will fall with increasing z, until at high z only the brightest objects are included in the counts. Also affecting the counts is the behaviour of the volume element with z (equation (6.24)). We can calculate the expected number of counts as a function of apparent magnitude m for the case where galaxies are conserved, and their luminosity does not change, by integrating equation (6.26) over z with n given by (6.28).

This provides a useful baseline against which to compare the observed count numbers. Figure 6.5 shows a compilation of galaxy number-magnitude counts in the B and K bands.

Both plots have the Euclidean slope at low magnitudes but flatten off at large magnitudes. The full curves show the relationship for an Einstein–de Sitter Universe, with no evolution of galaxies, for comparison. The observed counts in the B band exceed the calculated numbers by more than an order of magnitude for the faintest sources, whereas in the K band the discrepancy is much less marked. Cosmological models with $\Omega < 1$, or with $\Omega_\Lambda > 0$, have more volume at high redshifts, so the predicted count rate at faint magnitudes can be increased by a

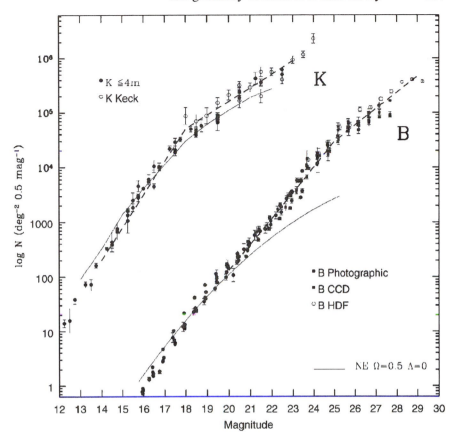

Figure 6.5. The figure shows number counts of galaxies in the B and K bands against magnitude. The full curve corresponds to non-evolving galaxies in an Einstein–de Sitter Universe. The broken curves are power-law fits to the data (Ellis 1997).

different choice of model. But a different model would not explain both curves. So galaxy evolution must play an important role in determining the shape of the curves. Pure luminosity evolution, which allows the luminosity of galaxies to increase with increasing redshift while conserving galaxy numbers, is the most straightforward case to consider. It raises the count rate at faint magnitudes, but again this does not seem able to account for the observed puzzling excess of faint blue galaxies (Ellis 1997) without contradicting the K-band results. The problem therefore seems to require a more complicated evolution of galaxies which must include allowance for the evolution of galaxy numbers as well as their luminosity. Suffice it to say that, at present, the problem of the faint blue galaxy excess is not fully understood. Until evolution is understood in detail it will not be possible to obtain cosmological parameters from the number-magnitude test.

6.6   The timescale test

We know that the age of the Universe must be greater than the ages of the stars and galaxies in it. Therefore, by determining the ages of the oldest objects we can put a lower limit on the age of the Universe. So the timescale test is a self-consistency test. If the ages of the oldest stars come out to be greater than the age of the Universe as determined from the parameters H_0, Ω_M and Ω_Λ then we have a so-called timescale problem. Such a state of affairs existed for a number of years until quite recently. Theoreticians favoured an $\Omega_0 = 1$ Universe with $\Omega_\Lambda = 0$, but globular cluster star ages were coming out to be $\sim$16 Gy. These could be reconciled only with a value of $H_0 \sim 50$ km s^{-1} Mpc^{-1} and not with the then favoured value of 80 km s^{-1} Mpc^{-1}. New developments have brought down both the ages of the stars and also the value of the Hubble constant, and there is now a better fit between the two age scales. We have shown how the age of a FLRW model Universe is calculated in chapter 5. Here we explain how the ages of astronomical objects are obtained and what limits they place on the three cosmological parameters.

6.6.1   The ages of the oldest stars

The primordial material from which the first stars were made consisted of about 75% hydrogen and 25% helium-4 by mass with traces of helium-3, deuterium and lithium-7. In chapter 7 we explain how these light elements were created in the early Universe and why the synthesis of heavier elements had to await the formation of stars. Thus the oldest stars are the ones with the lowest concentrations of metals in their envelopes, where metals in this context conventionally refers to elements of higher atomic weight than helium. This criterion identifies the stars in globular clusters as amongst the oldest in the galaxy. The globular clusters are dense spherical agglomerations of typically about 10^5 stars. The stars of a globular cluster appear to have formed at the same epoch. The amount of evolution that a particular cluster star has undergone will depend on the star's mass: the more massive a star the faster it evolves. This evolution can be depicted by the position of the star on a Hertzprung–Russell (H–R) diagram. In this diagram the two main observational quantities which characterize a star, its luminosity and surface temperature, are the y- and x-axes respectively of the plot but with the temperature increasing from right to left (see, for example, Zeilik *et al* 1997).

Young stars burn hydrogen to helium and are located on the main sequence of the H–R diagram: this is the heavily populated diagonal line running from low-luminosity and low-temperature stars at one end to high-luminosity, high-temperature stars at the other end. Stars start their lives on the main sequence and remain on it until they have used up the hydrogen in their cores. The position of a star on the main sequence is determined by its mass: the more massive a star is the hotter and brighter it is. Also the more massive a star is the faster it burns its hydrogen. So the most massive stars come to the end of their hydrogen-burning phase first. At this point the star starts to become more luminous, and also cools, so it moves off the main sequence and heads diagonally to the right along the red giant branch of the diagram. The lower the mass of the star the longer it spends on the main sequence. Thus, given that all the stars in a cluster formed at the same time, the mass of star that has just started to turn off the main sequence can be used to determine the age of the cluster. We get the mass of a star at the turn-off point by measuring its absolute visual magnitude. The theory of stellar evolution enables us to calculate the main sequence lifetime of such a star.

The main source of error in this method is in the determination of the distance to the globular cluster. If the distance is underestimated then the intrinsic luminosity of the stars comes out to be too faint and their mass comes out too small which, in turn, implies an age that is too great. Recent determinations of globular cluster distances have placed them further away than formerly, which has reduced their inferred ages.

The most recent detailed determination of globular cluster ages (Chaboyer *et al* 1998) gives an estimate for the mean age of 17 globular clusters of $11.5 \pm 1.3 \times 10^9$ years.

In order to estimate the present age of the Universe from this figure it is necessary to add on the age of the Universe at the time of the formation of the globular cluster stars. Allowing between 1 and 2 billion years gives an age for the Universe of between 12 and 15 billion years (Linweaver 1999). There is therefore no conflict between these ages and the age of the Universe obtained from the $m-z$ test described in section 6.4.

6.7 The lensed quasar test

The general theory of relativity predicts that when a ray of light passes near a concentration of mass, such as a galaxy, it suffers an angular deflection. It follows from this, as described in section 3.9, that if a galaxy lies between us and a distant quasar the light from the quasar will be lensed into a number of bright arcs. Lensing amplifies the intensity of the distant object so making it conspicuous.

The probability of a quasar being lensed by a galaxy along its line of sight depends on which cosmological model is assumed. The proper distance of a quasar having a redshift z is, from the metric equation (6.1),

$$D_{\mathrm{P}} = R_0 \int_0^r \frac{\mathrm{d}r}{(1 - kr^2)^{1/2}}.$$

Using equations (6.2) and (6.5) enables us to write this in terms of redshift z

$$D_{\mathrm{P}} = c \int_0^z \frac{\mathrm{d}z'}{H(z')}.$$

Finally, substituting for $H(z)$ from equation (6.9) gives the explicit expression

$$D_P = c \int_0^z \frac{dz'}{H_0[(1+z')^2(1+z'\Omega_M) - z'(2+z')\Omega_\Lambda]^{1/2}}. \tag{6.30}$$

Thus increasing Ω_Λ reduces the size of the denominator in (6.30) which increases the proper distance D_P. Increasing the distance to the quasars places more galaxies between them and us, so the probability of lensing is increased.

Observation of quasars has found only a few lenses among hundreds of quasars. Until recently this small number of lensed quasars was interpreted as evidence against a significant contribution to Ω_0 from a cosmological constant (Kochaneck 1996). However, a new analysis based on revised knowledge of E/S0 galaxies, which are thought to be principally responsible for the lensing, has changed this conclusion and it now seems that the low incidence of lensing events is consistent with $\Omega_M = 0.3^{+0.2}_{-0.1}$ and $\Omega_M + \Omega_\Lambda = 1$ (Chiba and Yoshii 1999). Thus this test is now in essential agreement with the magnitude–redshift test using SNe Ia .

6.8   Problems with big-bang cosmology

Having so far concentrated on the successes of the standard cosmology based on the FLRW models we turn now to some shortcomings of these models. To be clear, these are not so much failures, which would rule out the models, but incompletenesses and coincidences which might be the whole story but would preferably not be.

6.8.1   The horizon problem

In chapter 4 we saw that the microwave background is uniform in temperature across the sky to about one part in 10^5 after the dipole has been subtracted off. It seems reasonable to conclude that the explanation for this uniform temperature distribution is that the volume within our particle horizon must have come into thermal equilibrium prior to the epoch of last scattering when the background radiation set out on its journey to us. Now, in order to have attained thermal equilibrium, there must have been time for energy to flow from hotter to colder regions throughout our horizon volume so as to smooth out any initial temperature irregularities. The problem is that in the FLRW models this could not have happened! Regions of the sky separated by more than $\sim 2°$ had not been in causal contact with each other at the time of last scattering; so how could the whole 4π steradians of the sky have thermalized? This is the horizon problem.

We now prove these assertions. Consider, for simplicity, an Einstein–de Sitter Universe. The distance to the horizon at the epoch of last scattering is given by

$$D_h = 3ct_{ls} = 3ct_0(1+z)^{-3/2}$$

(equation (5.82)). So, as the geometry is Euclidean, this horizon length subtends an angle on the microwave sky given by

$$\Delta\phi = \frac{D_h}{rR} = (1+z)^{-1/2}[1-(1+z)^{-1/2}]^{-1}, \tag{6.31}$$

where rR is the proper distance to the surface of last scattering, which at redshift z is given by equation (5.76). At last scattering $z \sim 1000$, so (6.31) gives $\Delta\phi \sim 2°$. Thus areas of the microwave sky more than $2°$ apart lie outside each other's horizons so these regions could not have achieved the same temperature by exchanging radiation. The only answer that the standard model can provide is that the uniform temperature was an initial condition of the big bang. This is a rather lame answer, and it would be much more satisfactory if we could find some physical mechanism to circumvent the horizon problem and allow thermal equilibrium to be achieved. We shall show (later and in chapter 8) that the inflationary hypothesis provides such a mechanism and it also provides a natural resolution of the flatness problem which we consider in the next section.

6.8.2 The flatness problem

We have seen that our Universe has a density very close to the critical value, that is $\Omega_0 \simeq 1$. An exactly critical density Universe remains so at all subsequent times, but any small departure from the critical value at early times grows rapidly with the expansion of the Universe. To have a nearly critical density Universe at the present time requires that Ω be remarkably close to unity at the Planck time (defined in problem 5), the earliest time to which the Einstein equations can be pushed back. Thus our Universe seems to have required very finely tuned initial conditions. We are faced here with a choice similar to that we faced with the horizon problem: we can either assume special initial conditions or look for a mechanism that naturally produces the required fine tuning and explains the observation.

To illustrate this let us see how Ω evolves as a function of the scale factor $R(t)$. Let

$$\rho R^\alpha = \rho_0 R_0^\alpha, \tag{6.32}$$

so both the matter-dominated and radiation-dominated cases are covered by an appropriate choice of α. Using the definition of the density parameter Ω (5.32) to eliminate ρ and the Friedmann equation (5.30) as in section 5.12.1 we obtain

$$\frac{1-\Omega}{\Omega R^{\alpha-2}} = \frac{1-\Omega_0}{\Omega_0 R_0^{\alpha-2}}, \tag{6.33}$$

which, after rearrangement, yields

$$\Omega_0 = \left[1 + \left(\frac{R_0}{R}\right)^{\alpha-2} \frac{1-\Omega}{\Omega}\right]^{-1}. \tag{6.34}$$

Equation (6.34) shows that if $\Omega = 1$ then $\Omega_0 = 1$. But if we have a near critical density at the present time, $\Omega_0 \simeq 1$, then at earlier times Ω must satisfy

$$\left(\frac{R_0}{R}\right)^{\alpha-2} \frac{|1-\Omega|}{\Omega} \lesssim 1.$$

If $\alpha > 2$ and $t \ll t_0$ then $(R_0/R)^{\alpha-2}$ is a very large number, so $|1-\Omega|/\Omega$ must be a very small number. To show why this is a problem consider the evolution of Ω from the Planck time to the present. We need the value of R_0/R_P where R_P is the scale factor at the Planck time. Most of the expansion takes place in the radiation-dominated phase so we may take

$$\frac{R_0}{R_P} \sim \left(\frac{t_0}{t_P}\right)^{1/2} \sim 10^{30},$$

as $t_0 \simeq 3 \times 10^{17}$ s and $t_p \simeq 10^{-43}$ s. (During the matter-dominated phase R changes by only a factor of $\sim 10^4$ so the different dependence of R on t makes no essential difference to our conclusion.) So with $\alpha = 4$ in the radiation phase, to get $\Omega_0 \sim 1$ requires that $(1-\Omega_p)/\Omega_p \sim 10^{-60}$ or $\Omega_p = 1 \pm 10^{-60}$. If Ω differs from 1 by somewhat more than $\pm 10^{-60}$ then either Ω is driven to zero, in which case no structure can form, or Ω is driven to infinity and the Universe recollapses to a big crunch at an early stage. We conclude that this very stringent condition on Ω at the Planck time is required in order for a Universe in which stars, galaxies and life can appear.

6.8.3 The age problem

The early recollapse of the Universe in the absence of a fine-tuning of the density parameter provides an alternative expression of the flatness problem. It is remarkable that the Universe lasts long enough for stars to go through the cycles of evolution required to produce us (or alternatively that the parameters of atomic physics are so arranged as to produce us before the Universe recollapses or becomes too tenuous to produce galaxies). If Ω is not about 1 today, then the Universe cannot contain stars and have the right age, so the flatness problem can also be regarded as an aspect of the age problem.

6.8.4 The singularity problem

The explicit solutions for dust and for the radiation model in chapter 5 show that these have singularities in the past where the matter density tends to infinity. Another way of looking at this, which chimes better with the technical definition of a singularity, is that the path of any fundamental observer cannot be extended back into the past beyond a certain point, which we call the big bang. Thus the world lines of fundamental observers come into existence, along with time itself, not only for no apparent reason, but in such a way that no question of a reason can arise in physical terms. In general, for any form of normal matter, not just dust and radiation, the FLRW models must have singularities in the past. The acceleration equation gives

$$\frac{\ddot{R}}{R} = -\frac{4\pi G}{3}\left(\rho + 3\frac{p}{c^2}\right),$$

so if $\rho + 3p/c^2 \geq 0$, which is called the *strong energy condition* and is fulfilled for normal matter, then $\ddot{R} < 0$. Then, looking backwards in time, $\dot{R}$ is increasing in magnitude, thus R reaches zero in a shorter time than if $\dot{R}$ had remained constant. For constant $\dot{R} = H_0 R_0$, $R(t)$ reaches zero at $t = H_0^{-1}$. Hence $R = 0$ at some finite time $< H_0^{-1}$ ago (problem 51). That this is a real singularity at which $\rho \to \infty$ can be shown from the equation of local energy conservation in the form

$$\dot{\rho} = -3\left(\rho + \frac{p}{c^2}\right)\frac{\dot{R}}{R}.$$

This shows that $\dot{\rho}/\rho \leq -3\dot{R}/R$, hence that $\rho > \rho_0 R_0^3/R^3$ and therefore $\rho \to \infty$ as $R \to 0$.

Finally, the singularity theorems of Penrose and Hawking show that the existence of a singularity is not an accident of the symmetry of the FLRW models (Tipler *et al* 1980). Simplifying greatly, this is because either the microwave background itself, or the matter content, provide enough gravity to ensure that light-rays traced back into the past will start to reconverge before the last scattering surface, which is sufficient to ensure a singularity.

The obvious response to singularities of infinite matter density is that they signal the need to go beyond classical physics in the early Universe and that they will be dealt with in a consistent quantum theory of gravity.

6.9 Alternative cosmologies

Until recently the execution of the classical tests were insufficiently precise to determine the cosmological parameters. We have seen in this chapter that this is no longer the case; not only are some of the tests yielding values for Ω_M and Ω_Λ but these values are, at the time of writing, consistent with each other and also with values of the cosmological parameters obtained by other means. However it is almost certainly too early to claim that we have arrived at 'the correct' model of the Universe: the concordance that we now have could, with further developments, go away. In the past, supposed inconsistencies in relativistic models have led to searches for alternative theories: recall how the steady-state theory arose from apparent problems with the age of the Universe. The classical tests therefore have a second function: they can be used to good effect to rule out *ad hoc* cosmological models, provided only that one admits a modest knowledge of physics as a constraint on such speculative edifices.

We have already discussed the steady-state model in section 4.3.1. As a further example, we consider the 'tired light' hypothesis first proposed by MacMillan in the 1920s (Kragh 1996). One imagines a static Euclidean Universe in which light loses energy on its journey to us from distant galaxies due to some as yet unexplained physical process. It is not too difficult to account for Hubble's law in this theory. We can postulate

$$\frac{dE}{E} = \frac{d\nu}{\nu} = -H_0 \frac{dr}{c}$$

as the relation between frequency change $d\nu$ and distance dr. On integration, we get

$$r = (c/H_0) \log(\nu_e/\nu_0) = (c/H_0) \log(1 + z) = cz/H_0 + \cdots. \qquad (6.35)$$

Thus, for small z, redshift is proportional to distance, which is Hubble's law.

It is now easy to show that, despite this promising start, there is a conflict with the results of Kellerman for the angular diameter test (see figure 6.2). For, using (6.35),

$$\Delta\phi = d/r = (d/c)H_0 \log(1 + z).$$

Clearly the angle subtended by a standard rod as a function of z does not go through a minimum as found by Kellerman and as predicted for an expanding FLRW model Universe.

Other observations also rule out the simple tired light model. We saw in section 5.10 that the models based on the Robertson–Walker metric predict that, in addition to the redshifting of its light, all physical processes in a source at redshift z are observed to be slowed down, that is they are time dilated, by a factor $(1 + z)$. This effect has been observed in distant SNe Ia. See, for example, Riess *et al* (1997) and Liebundgut *et al* (1996). However, the tired light hypothesis is inconsistent with time dilation, since it predicts only the redshifting of light as it travels through space, with no affect on the arrival rate of photons. Finally, if photons were to lose energy as they travel through space, but their number density in a non-expanding Euclidean space were to remain constant, then a blackbody spectrum would not be preserved. Thus the tired light idea cannot explain either the origin or the blackbody character of the cosmic background radiation.

6.10   Problems

Problem 42. *Derive the Mattig relation, equation (6.7).*

Problem 43. *Derive the expression for $H(z)$ given in equation (6.9). Hint: Follow the derivation of equation (5.40) but include the Λ term in the Friedmann equation. Also recall that $\Omega_\lambda = \Lambda/3H^2$.*

Problem 44. (a) *For the Einstein–de Sitter model show that equation (6.12) reduces to*

$$\Delta\phi = d(1+z)^{3/2}\left(\frac{H_0}{2c}\right)[(1+z)^{1/2}-1]^{-1}.$$

(b) *Show that there is a minimum value for $\Delta\phi$ in the Einstein–de Sitter model at $z = 5/4$.*

Problem 45. *Estimate the angular size of a galaxy at a redshift of $z = 1$. Assume that the diameter of a galaxy is 20 kpc. Hint: Do the calculation for an Einstein–de Sitter Universe.*

Problem 46. *Show that the K correction takes the form*

$$K(z) = 2.5(1-\alpha)\log(1+z)$$

for a source spectrum of the form Radiated Power $\propto \nu^{-\alpha}$.

Problem 47. (a) *Consider a Universe with several components of mass-energy. Let ρ_i be the mass-density of the ith component and $p_i = w_i\rho_i c^2$ the pressure of this component. Show that the deceleration parameter $q = -\ddot{R}R/\dot{R}^2$ is*

$$q = \frac{\Omega}{2} + \frac{3}{2}\sum_i w_i\Omega_i,$$

where Ω is the total density parameter. Hint: Use the acceleration equation.

(b) *For a critical density Universe containing pressureless matter and a cosmological constant show that*

$$q = \frac{1}{2} - \frac{3}{2}\frac{\Omega_\Lambda}{[(1+z)^3\Omega_M + \Omega_\Lambda]}$$

where $\Omega_\Lambda = \Lambda/3H_0^2$, $\Omega_M = 8\pi G\rho_M/3H_0^2$. Hint: Use the expression for H given in equation (6.9) with the condition $\Omega_M + \Omega_\Lambda = 1$.

(c) *Hence show that this Universe goes from deceleration to acceleration at a redshift*

$$1+z = \left(\frac{2\Omega_\Lambda}{\Omega_M}\right)^{1/3}.$$

Problem 48. *For the Einstein–de Sitter model show that the volume of space V as a function of redshift satisfies*

$$\frac{dV}{dz} = 2\pi\left(\frac{2c}{H_0}\right)^3\frac{[1-(1+z)^{-1/2}]^2}{(1+z)^{9/2}},$$

and that dV/dz has a maximum at $z = 40/81$.

Problem 49. (a) Show that for $z \ll 1$ the expression for the volume of space in the redshift range dz,

$$\frac{dV}{dz} = 2\pi \left(\frac{2c}{H_0}\right)^3 \frac{[\Omega_0 z + (\Omega_0 - 2)\{-1 + (1 + \Omega_0 z)^{1/2}\}]^2}{\Omega_0^4 (1 + z)^6 (1 + \Omega_0 z)^{1/2}}$$

reduces to

$$\frac{dV}{dz} = 4\pi \left(\frac{c}{H_0}\right)^3 z^2,$$

and find the value of z at which dV/dz changes sign.

(b) Use equation (6.25) to obtain dN/dz for an Einstein–de Sitter Universe with the galaxy number conserved and find the value of z for which dN/dz is a maximum.

Problem 50. Show that in a very low density Universe ($\rho \approx 0$)

$$r_e R_0 = \frac{c}{H_0} \frac{z(1 + z/2)}{1 + z}.$$

Sketch the Hubble plot.

Problem 51. By expanding $R(t)$ as a Taylor series to second order in $t - t_0$, show that if $\rho + 3p/c^2 > 0$ then $R(t) = 0$ for some time $t_0 - t$ in the past less than H_0^{-1} ago.

Problem 52. It is proposed that the redshift of galaxies can be accounted for in a static Euclidean space if the masses of elementary particles change with time. Show that Hubble's law can be derived from the Bohr theory of the atom if the mass of a particle satisfies

$$m(t) = \frac{m R(t)}{R(t_0)},$$

where m is a constant and $R(t)$ is a time-dependent scale factor. Deduce that masses were smaller in the past in this theory. More exactly, show that we have

$$\frac{1}{1 + z} = 1 - \frac{H_0 r}{c} - \frac{1}{2} q_0 \left(\frac{H_0 r}{c}\right)^2 + \cdots,$$

where $r = c(t - t_0)$ is the Euclidean distance of a source having redshift z and $H_0 = (\dot{R}/R)_0$.

Suppose now that the luminosity of a galaxy L is related to its mass M by $L \propto M^\beta$. Obtain a relation between the number counts of galaxies $\log N(> S)$ brighter than a given flux, S in the form

$$\log N(> S) = -\tfrac{3}{2} \log S - \text{constant} \times (\beta + 1) S^{-1/2} + \cdots.$$

Observations show that this relation is flatter for distant sources than the Euclidean value $N \propto S^{-3/2}$. Show that this requires galaxies to be more luminous in the past ($\beta < -1$). (This conflicts with stellar evolution theory, which tells us that the luminosities of stars decrease with decreasing mass. In this theory it would also be difficult to explain the minimum in the apparent diameter of galaxies as a function of redshift.)

Chapter 7

Hot big bang

7.1 Introduction

The history of the Universe can be divided into three chapters. About the most recent phase we have much information and little certainty. This is the era when matter dominates the background radiation and complex material structures form and evolve. We shall discuss it in chapter 9. About the earliest phase we have no information at all, except that it was there. What theory we have leads to great speculations which will form the subject matter of chapter 8. The simplest and best understood phase is the period in which radiation and matter are in equilibrium with the radiation density dominating the matter density. During this time the Universe is filled with a uniform mixture of radiation and matter under conditions for which the physics is well understood. This is the subject of the present chapter.

In chapter 4 we learnt that the Universe contains radiation having a blackbody spectrum at a temperature $T_0 = 2.725$ K. There we showed that the effect of the expansion of the Universe on the radiation is to preserve the blackbody spectrum, but to lower the temperature. Looking back to a redshift z the temperature of the radiation is given by

$$T = T_0(1 + z).$$

The energy density of the radiation increases as $(1 + z)^4$ and the energy density of matter increases as $(1 + z)^3$; it follows that at redshifts greater than z_{eq}, given by

$$1 + z_{eq} = 2.4 \times 10^4 \Omega_M h^2,$$

the equivalent mass density of the radiation and the background neutrinos was greater than that of matter (see section 7.4 and problem 58). Recall also, as explained in section 5.14.1, that the effects of curvature and of a cosmological constant are negligible at these early times. Consequently the dynamics of the expansion before z_{eq} is determined by the mass density of the radiation alone. In this case we saw in section 5.16 that the scale factor evolves with time as

$$R(t) \propto (1+z)^{-1} \propto t^{1/2}.$$

An important consequence is that in this epoch temperature and time are uniquely related. To set the scale it is useful to remember that, to order of magnitude, an energy (kT) of 1 MeV or a temperature of 10^{10} K corresponds to a time of 1 s.

At high enough temperatures matter is again important, in a sense, because particle–antiparticle pairs are created by photon interactions. For example, for temperatures $kT > m_e c^2 \sim 0.5$ MeV photon energies exceed the electron rest mass-energy $m_e c^2$, and the electromagnetic interaction

$$\gamma + \gamma \rightleftharpoons e^- + e^+ \tag{7.1}$$

creates electron positron pairs and maintains them in thermal equilibrium with the photons. Similarly, the weak interaction

$$\nu + \bar{\nu} \rightleftharpoons e^+ + e^-$$

maintains an equilibrium between electron–positron pairs and neutrino pairs. Provided the reactions are fast enough (compared with the rate at which conditions are changing because of the expansion of the Universe) the plasma will remain in thermal equilibrium. For simplicity, we assume that the matter component is either ultra-relativistic $(kT \gg mc^2)$, in which case we neglect the particle rest mass m or we assume the matter is non-relativistic $(kT \ll mc^2)$, in which case this interaction is negligible anyway. At the temperatures and densities prevailing in the early Universe relativistic matter and radiation can be treated as perfect gases.

At early enough times all the currently known quarks and leptons will be present. As the plasma cools the equilibrium abundances in equation (7.1), and analogous reactions, shift to favour the photons and neutrino pairs over the massive particle species. With declining temperature the particles and their partner antiparticles drop out of equilibrium one by one, effectively annihilating into neutrinos and photons. Note that the neutrino–antineutrino pairs do not annihilate precisely because the weak interactions between them become too weak to bring this about in the lifetime of the Universe, and they are therefore still present as an undetectable background.

After about 10^{-4} s in the history of time, when the temperature was about 10^{12} K, the physics is understood and accurate calculations can be carried out. We can pick up the story at this time without reference to what preceded it, because equilibrium states carry no memory of how they arose.

As the Universe cools further to below about 10^9 K nuclear fusion reactions lead to the synthesis of nuclei of deuterium, helium and lithium. Nucleosynthesis stops at about 10^8 K partly because of the difficulty of overcoming the Coulomb barriers to interaction in a relatively cool plasma, but mainly because of the effect of a rapidly falling density on the relevant reaction timescales. The heavier elements can be made in stars, but the abundance of ^{4}He, which is about 25% by mass of the baryonic matter in the Universe, is far too large to be the result of stellar nucleosynthesis (problem 54). This account of the origin of the light elements is one of the major triumphs of the hot big-bang theory.

At times prior to 10^{-10} s the prevailing temperatures and densities reach realms not reproducible in the laboratory. Our current understanding of particle physics (called the *standard model*) enables us to explore conditions at least back to 10^{-10} s (corresponding to electroweak unification), and, with reasonable confidence, back to 10^{-35} s. At earlier times we reach the period of the *very early Universe* where we must resort to more speculative grand unified theories which at the present time lack experimental support. Finally, when we reach 10^{-43} s, the temperature is about 10^{31} K and the de Broglie wavelength of a typical particle is less than its Schwarzschild radius. At this point quantum effects are significant and general relativity, the classical theory of gravity on which our discussion has been based, ceases to be valid. This will be the era of quantum gravity.

In the sections that follow we look at the various epochs in the evolution of the Universe, along with the physics relevant at these times.

7.2   Equilibrium thermodynamics

It is important to realize that in the early Universe the particles can be treated as effectively non-interacting (as in a perfect gas; see problem 53). Without this simplification the physics of the early Universe would be intractable. For a perfect gas the dependence of energy density on temperature goes as follows. Bosons, integer spin particles such as photons and gauge particles, have a Planck spectrum for which the number of particles per unit volume having energies between E and $E + dE$ is given by

$$n(E)\, dE = \frac{4\pi g_B}{h^3} \frac{E^2\, dE}{c^3} \frac{1}{e^{E/kT} - 1},$$

where g_B is the number of internal degrees of freedom. For example $g_B = 2$ for photons because there are two states of polarization.

In the ideal gas approximation fermions, spin-$\frac{1}{2}$ particles such as electrons, quarks, protons and neutrinos, have a spectrum

$$n(E)\, dE = \frac{4\pi g_F}{h^3} \frac{E^2\, dE}{c^3} \frac{1}{e^{E/kT} + 1}, \tag{7.2}$$

with g_F again the number of internal degrees of freedom. For example, $g_F = 1$ for each neutrino and for each antineutrino (because neutrinos have only left-handed polarization states) and $g_F = 2$ for electrons and positrons because they each have two spin states. Note that we have neglected the chemical potential (or Fermi energy) which would otherwise appear in (7.2) (see appendix A, p 154). For the electron component this follows from the smallness of the η parameter which we consider in section 7.8. Strictly, for neutrinos and antineutrinos the vanishing of the chemical potential is an assumption in the standard hot big-bang picture.

At any time in the early Universe the particle species present are determined roughly as follows. Charged particles with rest energy $mc^2 < kT$ will be created through the reaction

$$\gamma + \gamma \rightarrow \text{particle} + \text{antiparticle,}$$

and will be maintained in thermal equilibrium with the photons through the reverse reaction in which the particle and antiparticle annihilate into photons. The number densities of the particle species are given by (7.2) and are comparable to the number density of photons.

Once the temperature falls below the pair production threshhold for a given species, i.e. for $mc^2 > kT$, that species will annihilate into photons and effectively disappear. On the other hand, if the rate of the reaction maintaining a particle species in equilibrium falls below the rate of expansion before the pair production threshhold is reached, then annihilation cannot occur and the particle species will remain subsequently in free expansion. Non-interacting relativistic particles generally have a temperature below that of the photons because the annihilation of other particle species transfers entropy to the photons thereby raising the temperature of the photons above that of the freely expanding particles. We shall explore the details of this for the case of neutrinos in section 7.7.3.

The corresponding overall energy density of the bosons is

$$u = \int_0^\infty En(E)\,\mathrm{d}E = \left(\frac{g_\mathrm{B}}{2}\right) aT^4$$

(see appendix A, p 156). Similarly, for the fermions, in the relativistic limit $E \approx cp$, the overall energy density is

$$u = \frac{7}{8}\left(\frac{g_\mathrm{F}}{2}\right) aT^4.$$

For both bosons and fermions the corresponding pressures are $p = u/3$.

For an assembly of particle types we sum over the constituents to get the overall energy density. This is written as

$$u = \frac{g_*}{2} aT^4,$$

where g_* is the effective number of degrees of freedom and is given by

$$g_* = \sum_i g_{\mathrm{B}i} + \frac{7}{8}\sum_i g_{\mathrm{F}i}, \tag{7.3}$$

the sums being taken over all types of bosons and all types of fermions respectively. If a species is expanding freely, hence at a temperature $T_i \neq T$, then the energy density of the species acquires a factor $(T_i/T)^4$.

When there are a large number of interacting species, at very early times and at high temperatures, g_* is large. For $kT \gg 300$ GeV all the particles and antiparticles of the standard model are present in equilibrium with photons and $g_* = 106.75$ (Kolb and Turner 1990). At the opposite extreme, below about 1 MeV, when electron–positron pairs are no longer present and the only relativistic species apart from photons are the three generations of (massless) neutrinos, g_* drops to 3.36. This is, therefore, the value of g_* at the present time and the energy density for photons and massless neutrinos is accordingly $1.68aT^4$.

7.2.1 Evolution of temperature: relativistic particles

How are these results modified in an expanding Universe? In chapter 4 we saw that, in an expanding Universe filled only with blackbody radiation, the Planck spectrum is preserved on expansion but that the temperature drops as $T \propto 1/R$ and the number density of photons as $1/R^3$. A similar result holds for massless fermions, or to a sufficient approximation, for ultrarelativistic massive fermions. The argument is exactly parallel to that for photons (see problem 18).

7.2.2 Evolution of temperature: non-relativistic particles

At the opposite extreme what can we say about a Universe filled with non-relativistic matter in thermal equilibrium at a non-zero temperature T? This will be important once the Universe has cooled sufficiently. In this case the matter has a Maxwell–Boltzmann distribution such that the number of particles with velocity between v and $v + dv$ is

$$n(v)\, dv = 4\pi N \left(\frac{m}{2\pi kT}\right)^{3/2} v^2 e^{-mv^2/2kT}\, dv, \qquad (7.4)$$

where N is the total number of particles in a comoving volume.

To see how this changes with time we need to look at how the velocity v of a given particle changes as a result of expansion. This change occurs because the particle is always overtaking observers who are moving away from it as the Universe expands. Let observer O measure particle P to have speed $v(t)$ at time t. Let observer O′, a distance $dx = v\, dt$ from O, measure a velocity $v(t + dt)$ at time $t + dt$. The speed of O away from O′ is $H\, dx$. Therefore $dv = v(t + dt) - v(t) = -H\, dx = -\dot{R}v\, dt/R$. This gives us

$$\frac{dv}{v} = -\frac{dR}{R}.$$

Hence

$$v \propto \frac{1}{R(t)}.$$

Another way to see this is to use Liouville's theorem which states that the volume of phase space containing a fixed set of particles has a phase volume that is constant in time. This phase volume is $4\pi = \mathrm{d}V v^2 \,\mathrm{d}v$. Since $V \propto R^3$ then to keep $\mathrm{d}V v^2 \,\mathrm{d}v$ constant we must have $v \propto R^{-1}$.

We now use this in the Maxwellian velocity distribution in the same way that we used the change in photon frequency in the Planck distribution. Assume the particle distribution evolves to a Maxwell–Boltzmann distribution at temperature T' and follow the same set of particles. Let $n(v)\,\mathrm{d}v$ be the number density of particles with speeds in the range v to $v + \mathrm{d}v$. Since the total number of particles is constant, we get

$$n(v)\,\mathrm{d}v V = n(v')\,\mathrm{d}v' V'.$$

Hence

$$n(v)\mathrm{d}v = 4\pi n(v')\frac{V'}{V}\left(\frac{m}{2\pi kT'}\right)^{3/2} v'^2 e^{-mv'^2/2kT'}\,\mathrm{d}v'$$

$$= 4\pi n(v)\left(\frac{m}{2\pi kT'}\right)^{3/2}\left(\frac{vR}{R'}\right)^2 e^{-mv^2 R^2/2kT'R'^2}\,\mathrm{d}v R/R'$$

$$= 4\pi n(v)\left(\frac{mR^2}{2\pi kT'R'^2}\right)^{3/2} v^2 e^{-mv^2 R^2/2kT'R'^2}\,\mathrm{d}v,$$

where we have used $v' = vR/R'$. This is a Maxwell–Boltzmann distribution with temperature $T = T'R'^2/R^2$, i.e. $T \propto R^{-2}$. Thus the distribution of speeds in a non-relativistic gas remains Maxwellian but as the Universe expands the temperature decreases with $TR^2 = \text{constant}$.

Another way to look at this is to consider again a box of gas that expands with the Universe. Since the expansion can be taken to be adiabatic (homogeneity implies that any loss of energy to the surroundings is balanced by a gain from the surroundings) we have

$$TV^{\gamma-1} = \text{constant}$$

with $\gamma = 5/3$ for a monatomic gas. From this we again see that $T \propto R^{-2}$.

There are several potential sources of confusion here if we take the expanding box analogy too literally. In the laboratory the gas in a container that is expanding sufficiently rapidly will be driven away from thermal equilibrium. This will happen if the timescale associated with expansion, $R/\dot{R}$, is small compared to a collision time in the gas, because the gas near the expanding walls of the container will not have time to thermalize. The expansion will not then be reversible. In an expanding Universe the gas is not driven away from equilibrium by rapid expansion because the expansion occurs throughout the gas, not just at the walls of the container. On the other hand, in the laboratory the expansion must be fast enough if conditions are to remain adiabatic; too slow an expansion would allow heat transfer from the surroundings. In the Universe the surroundings are always at the same temperature as the imaginary box, since they undergo the same expansion, so an arbitrarily slow expansion is adiabatic. These arguments are valid for a Universe filled solely with radiation or with relativistic particles, as well as for the case of a non-relativistic gas we are considering here: the maintenance of thermal equilibrium in an expanding Universe in these cases is a geometrical effect that does not depend on collision times. The same is true for a non-interacting mixture of these components: each behaves independently of the others. If matter and radiation interact, as they do in the real Universe, the picture is very different and is considered in the following.

7.3   The plasma Universe

We have seen that the matter temperature and the radiation temperature behave differently as the Universe expands. Thus, if there were no interaction between matter and radiation, the two temperatures would differ at later times even if they were to agree initially. This is precisely the case at the present time. Intergalactic neutral hydrogen is virtually transparent to 3 K blackbody radiation. Even a hot ionized intergalactic plasma of the maximum density permitted by, for example, Lyman continuum absorption would allow a microwave photon a 99% chance of making it across the present Universe without being scattered by electrons (problem 55). Furthermore the random motions in clusters of galaxies are scarcely perturbed by the presence of the microwave photons! In the present Universe, therefore, the background radiation and the matter each behave as if the other were not there.

As we look back into the past, however, there comes a stage where the radiation temperature is sufficiently high to ensure almost complete ionization of matter. This occurs at around 10^3 K (see later). The 3 K background exceeds 10^3 K at a redshift of about 1000. This is well before the advent of galaxies so at this time the Universe was filled with a fully ionized plasma. The uniformity of the background radiation leads us to conclude that this plasma was close to homogeneous. Now, in fully ionized matter of sufficient density the radiation and matter interact strongly. According to what we stated earlier, if the timescale for this interaction is smaller than the expansion timescale then the matter and radiation will be able to come into equilibrium and remain so. Therefore at sufficiently early times, the matter temperature T_m and the radiation temperature T_r were equal.

It should now be obvious that this raises a problem. For if radiation and matter cool at different rates in the absence of interactions, how does their common temperature behave when they interact? The approximate answer is surprisingly simple: while matter and radiation are strongly coupled the radiation behaves as if the matter were not there and the common temperature follows that of the radiation. To see this consider the specific heats of the gas and of the radiation. To raise the temperature of unit mass of matter by ΔT_m requires an input of heat $C_m \Delta T_m$ where C_m is the appropriate specific heat per unit mass.

For an order of magnitude estimate we can take $C_m = C_V \sim k/m_p$. The total amount of internal energy of the radiation that could be available to heat unit mass of matter of mass density ρ_m is of order aT_r^4/ρ_m. If the extra heat loss of the faster-cooling matter is made up by extracting energy from the radiation, we have

$$-\Delta\left(\frac{aT_r^4}{\rho_m}\right) \sim \frac{k}{m_p}\Delta T_m.$$

Therefore, ignoring for the moment the contribution from non-baryonic matter,

$$\frac{\Delta T_r}{\Delta T_m} \sim -\frac{kn_b}{s_\gamma} = 0.28\eta,$$

where the final equality follows from equation (7.11). So imagine that the matter temperature has fallen significantly below the radiation temperature and is raised back up again. Then $\Delta T_m \sim T_r$, and $\Delta T_r/T_r \sim 0.28\eta = 7.3 \times 10^{-9}\Omega_B h^2$. Thus, in the process of reheating the matter, the radiation temperature changes by no more than one part in 10^8. We conclude that, provided the interactions exist to bring about equilibrium, the expanding mixture of radiation and matter in the proportions we find in our Universe, will cool like radiation alone as the Universe expands. We see that the matter is receiving energy from the radiation on the interaction time scale which is much shorter than the expansion timescale, so it is not free to respond to expansion by cooling adiabatically as $T_m \propto R^{-2}$. On the other hand, weakly interacting dark matter will decouple from radiation at an early stage and will therefore cool independently.

7.4 The matter era

The densities of matter and radiation, including neutrinos, are equal at a temperature T_{eq} which we can calculate as follows. The radiation density obtained from the present temperature of the microwave background is

$$\rho_r = \left(\frac{g_*}{2}\right)\frac{a}{c^2}T^4 = 8.09 \times 10^{-31}\left(\frac{T}{T_0}\right)^4 \text{ kg m}^{-3}.$$

Since $T \propto 1/R(t)$ and $\rho \propto R^{-3}$ the matter density is given by

$$\rho_m = 1.88 \times 10^{-26}\Omega_M h^2 \left(\frac{T}{T_0}\right)^3 \text{ kg m}^{-3},$$

where T is again the radiation temperature, not the matter temperature. These are equal when

$$T_{eq} = 2.4 \times 10^4 T_0 \Omega_M h^2 \text{ K}$$

or about 9700 K with $\Omega_M = 0.35$ and $h = 0.65$ (section 5.16). The corresponding redshift obtained from $1 + z = T/T_0$ is $z_{eq} = 2.4 \times 10^4 \Omega_M h^2 = 3000$, as we stated in the introduction.

These calculations are exact. To obtain the *time* at which the matter and radiation densities are equal we have to solve the evolution equations for a model containing both radiation and matter in the observed ratio. We can get an *estimate* if we take the evolution to be determined approximately by the matter content alone back to the point where the matter and radiation densities become equal. Then at early times, where $1 + z \gg \Omega_0^{-1}$,

$$t \sim \tfrac{2}{3}(1 + z)^{-3/2} H_0^{-1} \Omega_0^{-1/2} \tag{7.5}$$

gives the approximate conversion from redshift to time (problem 31). Hence putting, say, $\Omega_0 = \Omega_M = 0.35$ and $h = 0.65$, $t_{eq} \sim 5 \times 10^4$ years. An alternative estimate can be obtained if we evolve the Universe forward in time in the radiation era using equation (7.6) from the next section. An exact result can be derived in a model that contains both radiation and matter, since such a model gives the relation between redshift and time consistently. This calculation gives

$$t_{eq} = 1.03 \times 10^3 (\Omega_M h^2)^{-2} \text{ years}$$

(problem 59). This time represents the transition from a Universe in which the dynamics is dominated by the radiation content to a matter-dominated Universe.

7.5   The radiation era

7.5.1   Temperature and time

In the radiation-dominated era (section 5.16) we neglect the density of massive particles so

$$\rho = \rho_r = \frac{u_r}{c^2}$$

where u_r contains contributions from all forms of relativistic particles. Thus

$$\rho_r = \frac{g_*}{2c^2} a T^4.$$

But the density in the radiation era is close to critical (section 5.16) so

$$\rho_r = \frac{3H^2}{8\pi G} = \frac{3}{32\pi G t^2},$$

where $H = 1/2t$ in the radiation model, since $R \propto t^{1/2}$. Equating the two expressions for ρ_r gives

$$T = \left(\frac{3c^2}{16\pi G a g_*}\right)^{1/4} \frac{1}{t^{1/2}}$$

or

$$T = 1.6 g_*^{-1/4} t^{-1/2} \text{ MeV} \tag{7.6}$$

with t in seconds. Apart from the factor of g_* this is the same as the relation we found in section 5.16.1 for the radiation model. Thus, in the early Universe the temperature and time are closely related. For approximate estimates it is convenient to memorize this as $T = 10^{10}/t^{1/2}$, t in seconds, T in degrees Kelvin or, equivalently, $T \simeq t^{-1/2}$ with T in MeV and t in seconds, as we intimated in the introduction.

7.5.2 Timescales: the Gamow criterion

The effectiveness of the various processes that bring about the energy transfer between radiation and matter determines how close the radiation and matter temperatures are. Various eras in the evolution of the combined system can be identified in term of the dominant interaction mechanism as we shall discuss in the following sections. The presence of even three particle types (photons, electrons, protons) and one interaction (electromagnetism) leads to numerous possibilities. The introduction of the weak interaction and neutrinos at early times and neutral atoms at late times adds to the complexity. The basic idea, however, is simple: one is interested at each epoch in the interaction which has a timescale t_{int} of about the same order as the expansion timescale t_{exp}. This leads to the Gamov criterion: the time at which an interaction ceases to be effective is determined by the condition

$$t_{\text{int}} \lesssim t_{\text{exp}}.$$

Before these timescales become equal the interaction maintains equilibrium; subsequently it plays no role and the relevant particles decouple from the action.

It is sometimes useful to think of the Gamov criterion in another way. The expansion time scale is given by H^{-1} which is roughly equal to the age of the Universe at the time in question. For the radiation-dominated era we can relate the timescale to the temperature using the approximate relation we derived in section 5.16.1 or the more exact form from (7.6). For the matter-dominated era we use (7.5). For numerical estimates these give (e.g. Padmanabham 1993)

$$t_{\text{exp}} = 1.5 \times 10^{12} \left(\frac{T}{1 \text{ eV}} \right)^{-2} \text{ s} \qquad \text{for } t < t_{\text{eq}} \qquad (7.7)$$

$$= 1.1 \times 10^{12} (\Omega_0 h^2)^{-1/2} \left(\frac{T}{1 \text{ eV}} \right)^{-3/2} \text{ s} \qquad \text{for } t > t_{\text{eq}}.$$

The interaction timescale for particles with density n, speed v and interaction cross section σ is $t_{\text{int}} = (\sigma n v)^{-1}$. For relativistic particles this is close to $(\sigma n c)^{-1}$. The Gamov criterion becomes

$$\frac{t_{\text{exp}}}{t_{\text{int}}} = \frac{\sigma n c}{H} > 1.$$

But $c/H = d$ is the scale of the Universe (the Hubble distance) at time t. Thus the Gamow condition is

$$\tau = \sigma n d > 1, \qquad (7.8)$$

which states that significant interaction will occur if the optical depth τ, defined by the first equality in (7.8), across a Hubble distance d exceeds unity.

7.6   The era of equilibrium

If the radiation background were not thermal at the present time there would be no interactions now that would make it so. We can therefore ask what was the latest time at which thermalization could occur if the radiation were not exactly thermal at the beginning. This involves at least one subtlety. Consider interactions such as Compton scattering, which transfer energy between matter (electrons) and radiation but conserve photon number. We can then envisage a situation in which a *kinetic* equilibrium exists in which each transfer of energy is balanced by an equal but opposite exchange. In this case the spectral distribution of energy of both electrons and photons is preserved, but is not necessarily thermal. True thermal equilibrium requires a Planck spectrum for the radiation, hence a mechanism that can create or destroy photons as required. In other words, if the initial distortion of the spectrum involves photon number, thermal equilibrium can be restored only by interactions that generate photons. It turns out (Peebles 1971) that kinetic equilibrium (i.e. a balance of reaction rates) can be restored no later than the time corresponding to a redshift $z = 2.2 \times 10^4 (\Omega_B h^2)^{-1/2}$. But true thermal equilibrium can be established either by bremsstrahlung, the creation of a photon by the deceleration of a free electron in the field of an ion (which dominates if $\Omega_B h^2 > 0.1$) or double Compton scattering, the creation of a low-energy photon during Compton scattering, no later than

$$z \sim 2 \times 10^6,$$

omitting a factor weakly dependent on $\Omega_B h^2$.

7.7   The GUT era: baryogenesis

Above 10^{15} GeV, hence before 10^{-36} s, the strong and electroweak interactions appear to be unified, and mediated by massive particles, the X-bosons. If the decays of X and $\bar{X}$ into quarks and antiquarks occur at different rates, this provides the possibility of explaining the matter–antimatter asymmetry of the Universe. These statements are based on a number of facts and extrapolations.

The facts are as follows.

(1)   At energies above 1 TeV (10^{15} K) the weak and electromagnetic interactions have the same strength and are unified as aspects of a single electroweak force, similar to the way in which electric and magnetic forces are unified as the electromagnetic force. This unification predicted the existence of the vector bosons $W^{\pm}$ and Z^0 which were subsequently identified.

(2) The strengths of the three fundamental interactions, as measured by the effective coupling constants, appear to reach a common value around 10^{15} GeV. This, together with the success of the electroweak theory, suggests a further unification. This unification involves as yet unknown particles, generically called X, and their corresponding antiparticles $\bar{X}$, which can mediate the decay of quarks to leptons.

(3) The decays of the neutral mesons K^0 and $\bar{K}^0$ are asymmetric with respect to matter and antimatter. For example, K^0 is a mixture of a long-lived component K_L and a short-lived component K_S. The state K_L decays into $e^+ + \pi^- + \nu_e$ rather than $e^- + \pi^+ + \bar{\nu}_e$ about 503 times in 1000 decays. This is an example of the effect known as CP violation. (C refers to interchanging particles and antiparticles, P to interchanging left-handed and right-handed polarization states.)

(4) An equilibrium mixture of K_L and $\bar{K}_L$ will contain equal numbers of particles and antiparticles. To produce an imbalance requires in addition to violation of C or CP both the non-conservation of baryons (or B violation) and a non-equilibrium situation.

(5) Baryon number violation is provided by the X particles. For example X decays to two quarks (baryon number 2/3) or to a quark and a lepton (baryon number 1/3).

(6) Expansion in the early Universe, if rapid enough compared with the interaction timescale, provides a non-equilibrium environment.

The extrapolation involves postulating that the $X\bar{X}$ of grand unification behave like the $K\bar{K}$ with respect to symmetry under interchange of particles with antiparticles (C or CP violation). Then the decay rate of X to qq differs from that for the decay to $\bar{q}\bar{q}$ and all the conditions for baryogenesis can be satisfied. The simplest theory of this type predicts that the proton is unstable (for example, it can decay into $e^+ + \pi^0$) with a lifetime no more than 3×10^{31} years, which is ruled out by the observation that the proton half-life exceeds 1.6×10^{33} years (Shiozawa *et al* 1998). There is, however, no shortage of alternative unified theories.

7.7.1   The strong interaction era

Whatever the explanation, the Universe just below 10^{15} GeV contains a plasma of free quarks, antiquarks and gluons with of order $10^9 + 1$ quarks for every 10^9 antiquarks. The quarks pairs annihilate as kT falls below the pair production threshholds for the various quark flavours. Below 0.2 GeV the quarks, and any residual antiquarks, are no longer free but confined in hadrons (Schwarzschild 2000). The mesons and unstable baryons decay leaving roughly one baryon (proton or neutron) for every 10^9 photons.

7.7.2   The weak interaction era: neutrinos

The reactions that maintain the neutrinos in equilibrium are

$$\nu + \bar{\nu} \leftrightarrow e^+ + e^-$$

and elastic scattering of neutrinos on electrons. The cross sections σ for these reactions are of order $G_F^2(kT)^2(\hbar c)^2$ where $G_F \sim 1.2 \times 10^{-5}$ GeV^{-2} is the Fermi coupling constant for the weak interactions. The number density of interacting particles is $\sim \pi^2(kT)^3/(\hbar c)^3$ (appendix A, p 157). So the reaction rate is

$$\Gamma = n\sigma v = G_F^2(kT)^5/\hbar$$

with $v \sim c$, and the timescale is Γ^{-1}. Thus, using (7.7) in the radiation era,

$$\frac{\Gamma^{-1}}{t_{\text{exp}}} = \left(\frac{1.6 \times 10^{10} \text{ K}}{T}\right)^3.$$

The interaction ceases to keep up with the expansion when the reaction timescale Γ^{-1} exceeds the expansion timescale t_{exp}. Thus the neutrinos decouple below about 10^{10} K.

7.7.3   Entropy and $e^- - e^+$ pair annihilation

We have seen that at very high temperatures there are many particle–antiparticle species in thermal equilibrium. One by one these species annihilate into photons as kT falls below the respective rest energies. To a good approximation the total entropy is conserved during annihilation and the photons take up the entropy of the pairs. We now consider the last of these annihilation episodes, the conversion of electron–positron pairs to photons that occurs as the temperature drops below 0.15 MeV.

The cross section for annihilation is of order the Thomson cross section, $\sigma_T = 6.65 \times 10^{-29}$ m^2. The timescale for annihilation per electron is $\Gamma_{e^\pm}^{-1} = (\sigma_T n_e c)^{-1}$, where n_e is the electron density, and we take relativistic electrons to have a speed approximately equal to c. The electron density is $n_e \sim n_{\text{ph}} \sim (aT^3/k) \sim 10^{37}$ m^{-3} at $T \sim 10^{10}$ K. So the annihilation timescale is $\Gamma_{e^\pm}^{-1} \sim 10^{-17}$ s. This is much less than the expansion timescale at 1 s (which is equal to 1 s, of course). The annihilation timescale is also less than the expansion timescale above 10^{10} K, but then the reverse creation of $e^\pm$ pairs maintains the $e^\pm$ plasma in equilibrium. Once this is not possible the pairs are rapidly consumed by annihilation.

The annihilation energy goes entirely into the photons since the neutrinos are essentially decoupled by this time. This leads to a heating of the photons above the neutrinos. We need to calculate this heating since it affects the radiation entropy per baryon at the present. Assume the electron–photon plasma moves through a sequence of equilibrium states as annihilation proceeds so the annihilation is reversible. (The shortness of the annihilation timescale makes this inevitable.) At each infinitesimal step the heat energy, dQ, extracted from the $e^\pm$ annihilations in a comoving volume V of plasma, is supplied to the radiation at the same temperature. Therefore

$$dQ = -T \, dS_{e^\pm} = T \, dS_\gamma$$

and the entropy of $e^\pm$ pairs goes into photon entropy, S_γ. Before annihilation the three components are at the same temperature, $T_\gamma = T_{e^\pm} = T_\nu$, and, from (7.22),

$$S_{\gamma + e^\pm} = S_\gamma + S_{e^\pm} = \tfrac{4}{3} a T_\gamma^3 V + \tfrac{4}{3}(2 \times \tfrac{7}{8} a T_{e^\pm}^3) V = \tfrac{11}{3} a T_\gamma^3 V$$
$$= \tfrac{11}{4} S_\gamma.$$

After annihilation all this entropy is in the form of photons, so the new photon entropy is

$$S_\gamma' = \tfrac{11}{4} S_\gamma.$$

The photon temperature after annihilation, T_γ', is therefore given by

$$\frac{11}{4} = \frac{S_\gamma'}{S_\gamma} = \left(\frac{T_\gamma'}{T_\gamma}\right)^3 \left(\frac{V'}{V}\right). \tag{7.9}$$

Were the comoving volume not expanding during annihilation we could use (7.9) to make a direct statement about how the temperature of the radiation changes. To remove the ratio of volumes in (7.9) consider now the neutrinos. These are non-interacting, so their entropy in a comoving volume does not change, the temperature changing only as a result of expansion:

$$1 = \frac{S_{\nu\bar\nu}'}{S_{\nu\bar\nu}} = \left(\frac{T_\nu'}{T_\nu}\right)^3 \left(\frac{V'}{V}\right). \tag{7.10}$$

Thus, combining (7.9) and (7.10) we can say

$$\frac{T_\gamma'}{T_\nu'} = \left(\frac{11}{4}\right)^{1/3} \simeq 1.4.$$

The temperature of the radiation field is therefore increased by about 40% over that of the neutrinos, hence over what it would have been without electron–positron annihilation, as the Universe cools through this phase.

7.8 Photon-to-baryon ratio

The entropy density of blackbody radiation at temperature T is $s_\gamma = 4aT^3/3$ and the number density of photons is given by $n_\gamma = aT^3/(2.7k)$ (appendix A, p 157).

So the specific entropy density s_γ/k, and the photon number density are simply related by

$$\frac{s_\gamma}{k} = 3.6 n_\gamma,$$

and hence measure the same thing.

Now, since the annihilation of $e^\pm$ pairs, $RT =$ constant, and the number density of baryons n_b is proportional to R^{-3} as baryons are conserved. It follows that the ratios $(s_\gamma/k)/n_b$ and n_γ/n_b are independent of time (for earlier times see appendix C). It has become customary to define a parameter η by

$$\eta^{-1} = \frac{n_\gamma}{n_b} = 0.28 \frac{s_\gamma/k}{n_b}. \tag{7.11}$$

As we shall see in section 7.9 the ratio of photon number to baryon number, η^{-1}, is a useful parameter in nucleosynthesis calculations.

It will be helpful to express η in terms of the baryon density parameter at the present time Ω_B. Substituting for s_γ from (7.22) and using $n_B = \rho_B/m_p$, with $\rho_B = 1.88 \times 10^{-26}\Omega_B h^2$ (since $\rho_B = (\rho_c)_0\Omega_B$), gives

$$\eta = 2.7 \times 10^{-8} \left(\frac{2.7\ \mathrm{K}}{T_0}\right)^3 \Omega_B h^2. \tag{7.12}$$

Since the temperature of the microwave background is now well established, the parameter η is an alternative way of specifying the density parameter Ω_B, subject to uncertainties in the Hubble constant. We shall see in section 7.9.4 how a value for η (and hence Ω_B) can be obtained by comparing the observed abundance of deuterium with that deduced from the yield from cosmological nucleosynthesis as a function of η.

7.9   Nucleosynthesis

As the Universe cools below 10^9 K nuclear reactions become possible. Starting from a mixture of neutrons and protons heavier elements can be built up. For a complete picture all possible reactions of the light elements need to be considered in the temperature range 10^{11} K to below 10^9 K and the density range 10^4 kg m^{-3} to 10^{-2} kg m^{-3}. Although in detailed calculations elements up to oxygen are considered, the principal reactions are those shown in figure 7.1. The physics is standard, but depends on measurements of a large number of reaction cross sections. Most of the crucial ones in the figure are now known to sufficient accuracy; those that are not contribute about a 50% uncertainty in the abundance of ^{7}Li (Schramm and Turner 1998). Ultimately the yields have to be computed numerically from the coupled rate equations for nuclei in an expanding box filled with radiation, but we can get some insight into the results from the considerations in the following sections.

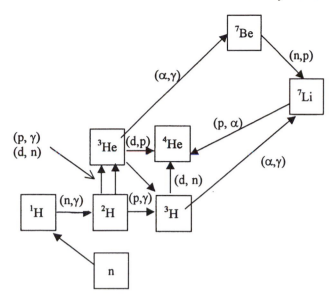

Figure 7.1. The main nuclear reaction network in the cosmological synthesis of the light elements.

7.9.1 Weak interactions: neutron freeze-out

The Universe at about 1 s contained photons, electron–positron pairs, protons and neutrons in thermal equilibrium at a temperature of about 10^{10} K, as well as neutrinos. The neutrinos have thermally decoupled from the e^+–e^- plasma, but they still interact inelastically to some extent with neutrons and protons converting between the two nucleon states according to the reactions

$$p + \bar{\nu}_e \rightarrow n + e^+ \qquad (7.13a)$$
$$n + \nu_e \rightarrow p + e^-. \qquad (7.13b)$$

This has a negligible effect on the abundant population of neutrinos, but is crucial for the far less abundant np population since it keeps them in equilibrium. The neutrino coupling to nucleons (protons and neutrons) via the weak interaction has a cross section about five times higher than that to leptons. This leads to decoupling of the nucleons at a slightly later time than the neutrino decoupling.

In the standard model it is assumed that the lepton number of the Universe is zero, i.e. that the numbers of neutrinos and antineutrinos are equal. A non-zero lepton number would imply a different balance between neutrons and protons that could be tuned to produce any amount of helium. We shall not consider this possibility, but note that it implies that zero lepton number is a key assumption of the standard model. In principle, neutrinos and antineutrino pairs can annihilate into $e^\pm$ pairs (via a Z^0) and hence into photons, but the interaction is so weak that the annihilation time is much greater than the Hubble time and so can be neglected.

The efficient interconversion of neutrons and protons at temperature T keeps them at their statistical equilibrium abundances given approximately by the Boltzmann formula

$$\frac{n_n}{n_p} = \exp\left[-\frac{(m_n - m_p)c^2}{kT}\right],$$

where n_n and n_p are the number densities of neutrons and protons and m_n and m_p are the masses of the neutron and proton. We have ignored the chemical potentials of the protons and neutrons, which is justified in appendix A, p 157. This interconversion becomes ineffective once the interaction timescale becomes longer than the expansion timescale, at time t_{np}, say. The second of the interactions (7.13b) is clearly related to the β-decay of the neutron, with mean lifetime $t_n = 887 \pm 2$ s, so we might guess t_{np} to be of order t_n. There are two factors that reduce this estimate significantly. One is that the reaction rates (7.13), unlike neutron decay, depend on the populations of neutrinos and electrons in the initial and final states. The other is that the reaction cross section is temperature dependent. The upshot is that the interaction timescale becomes equal to the expansion timescale at about $kT \sim 0.8$ MeV when

$$\frac{n_n}{n_p} \sim \frac{1}{6}.$$

The ratio is 'frozen in' at this value until the time t_n when β decay becomes significant for any remaining free neutrons. In fact, the freezing-in is not exact, since some neutrons will β-decay before t_n; by $kT = 0.3$ MeV the ratio declines slightly to 1/7, the value we shall use later (problem 61). Nevertheless, at this temperature the equilibrium ratio is 1/74, so the departure from equilibrium is significant.

7.9.2 Helium

While equilibrium holds the abundances of nuclei are small. This is because the large number of photons keeps the equilibrium:

$$\text{nucleus} + \text{photons} \rightleftharpoons \text{nucleons} \tag{7.14}$$

shifted to the right. In effect, nucleons are broken up as soon as they are formed by the abundant supply of photons. One might guess that nuclei would start to form once the temperature drops to $kT \sim B$, the nuclear binding energy per nucleon. Binding energies per nucleon range from 1.1 MeV for deuterium to 7.7 MeV for ^{12}C. In fact, nuclei are favoured only below about 0.3 MeV. To see this consider the simplest case of the deuterium nucleus. The dissociation into a neutron and proton by interaction with a photon is analogous to ionization of an atom, and the equilibrium abundances are governed by an equation of the same form, the Saha equation (appendix B). The equilibrium abundance of deuterium $n_D = (1-x)n_B$, at temperature T is given by

$$\frac{1-x}{x^2} = 3.8\eta \left(\frac{kT}{m_p c^2}\right)^{3/2} \exp\left(\frac{B}{kT}\right).$$

where B is the binding energy of deuterium and η^{-1} measures the radiation entropy (section 7.8). Since η is very small, and $kT/m_p c^2 \ll 1$, the deuterium abundance, $1 - x$, remains small until $kT \ll B$. At this point the exponential dependence of the nuclear abundances on binding energy eventually overcomes the photon dissociation. This occurs at energies some 30 times lower than the binding energy per nucleon. If the nuclei were able to remain in equilibrium at such a temperature they would rapidly build up into iron.

The Universe is saved from this fate by the rapid evolution which prevents equilibrium at this stage. Once kT reaches 0.1 MeV, the abundances of D($= {}^2$H) and ^{3}He become of order unity and the production of ^{4}He builds up rapidly. At this stage, at a time of around 3 min, essentially all the neutrons are incorporated into helium. The process goes no further because by this stage (1) the temperature is so low that further reactions are impeded by their Coulomb barriers; (2) the density is too low to build up ^{12}C by the triple α reaction (the fusion of three ^{4}He nuclei); and (3) there are no stable nuclei with mass numbers 5 and 8 which could act as intermediates. To a good approximation, therefore, all the neutrons are incorporated into helium nuclei, so there are half as many ^{4}He nuclei as there were neutrons and the number of H nuclei is the number of protons left over. The fraction of nuclei which are ^{4}He is

$$\frac{\text{number of } {}^4\text{He}}{\text{number of H} + \text{number of } {}^4\text{He}} = \frac{\frac{1}{2}n_n}{n_H + \frac{1}{2}n_n} = \frac{\frac{1}{2}n_n}{(n_p - 2 \times \frac{1}{2}n_n) + \frac{1}{2}n_n}$$

$$= \frac{\frac{n_n}{n_p}}{2 - \frac{n_n}{n_p}}.$$

If $n_n/n_p \sim 1/7$ this gives a fraction of about 7.5% helium by number. The relative abundance by mass, for which the standard designation is Y, is obtained by multiplying each number abundance by the corresponding particle mass (taking $m_n \approx m_p$),

$$Y = \frac{\text{mass } {}^4\text{He}}{\text{mass of } p + \text{mass of } n} = \frac{\frac{1}{2}n_n m_{{}^4\text{He}}}{n_p m_p + n_n m_p}$$

$$= \frac{2\frac{n_n}{n_p}}{1 + \frac{n_n}{n_p}} \sim 25\%$$

for $n_n/n_p \sim 1/7$. This is close to the observed abundance of helium.

7.9.3 Light elements

Because the binding energies of the light elements D ($\equiv$ ^{2}H), ^{3}H and ^{3}He are smaller than that of ^{4}He, the lighter nuclei are more easily broken up by photodisintegration (interactions with high-energy γ-ray photons) and their equilibrium abundances are small down to about $kT \sim 0.1$ MeV. At this point the equilibrium shifts to the left in equation (7.14) and significant abundances start to build up. However, at this stage ^{4}He has far less than its equilibrium value. This is because the supply of D and ^{3}He has not been able to keep up with requirements of the shift to favour nuclei that has already occurred for ^{4}He in equation (7.14). Thus, the production of ^{4}He rapidly consumes almost all the lighter elements as they are produced, making it impossible to amass their much larger equilibrium values. On the other hand, as we have seen, by the time the bottleneck has been broken for the production of helium, conditions are such that it is not possible to produce the next stable nucleus, ^{12}C. Thus the only element produced with a large abundance in the early Universe is ^{4}He. Nevertheless, as we shall see later, the trace amounts of D, ^{3}He and ^{7}Li that are produced turn out to be important probes of the conditions under which nucleosynthesis occurs.

Figure 7.2 shows the results of detailed computations (Schramm and Turner 1998) with a public version of Wagoner's code (Wagoner 1973, Kawano unpublished, see Kolb and Turner 1990). The results plot the computed abundances against the parameter $\Omega_B h^2$, or, equivalently, η (equation (7.12)).

It is a remarkable feature of the hot big-bang theory that it is able to reproduce abundances that are even approximately compatible with observations of both abundances and the microwave background temperature. (Although strictly there is no single value of η that is compatible with the claimed formal errors on the abundance measurements, uncertainties still exists so there is no contradiction between theory and observation.)

7.9.4 Abundances and cosmology

The abundances of the elements provide us with relics to probe the Universe between about 1 and 3 min, from neutrino decoupling to helium formation. We have seen in the previous section how the parameter η^{-1}, that is the photon-to-baryon ratio, is, in principle, determined by the relative abundances of the light elements. It is of interest to understand how this comes about and to investigate if there are any other cosmological parameters constrained by nucleosynthesis.

The abundance of helium depends on the neutron-to-proton ratio at freeze-out unless the baryon density is so very low that the Gamov criterion for the production of ^{4}He is not satisfied or so very high that the criterion for the conversion to carbon is. Neither of these are stringent requirements so the dependence of ^{4}He on η is weak. For larger values of η the abundances of D and ^{3}He build up somewhat earlier so ^{4}He production starts at a higher value of the neutron proton ratio n/p. This leads to somewhat more ^{4}He. The larger η

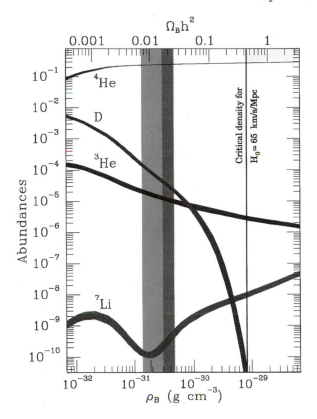

Figure 7.2. The mass fraction of ^{4}He and the number relative to hydrogen of deuterium (D), ^{3}He and ^{7}Li as a function of the present baryon density ρ_B (or, equivalently, $\eta = 2.7 \times 10^{-8}\Omega_B h^2$ from nucleosynthesis calculations. The dark band picks out the range of density consistent with the observed primeval deuterium abundance (Schramm and Turner 1998).

also means that D and ^{3}He are burnt rather more easily so the overall yield is decreased. This explains the general trend in figure 7.2. The result of comparison with observation is $\eta \simeq 5 \times 10^{-10}$ with the limits (Burles *et al* 1999)

$$0.017 \lesssim \Omega_B h^2 \lesssim 0.021.$$

On the other hand, ^{4}He is sensitive to g_* (equation (7.3)), hence to the number of relativistic particle species. This is because the rate of expansion at a given temperature depends on the number of particle species since

$$t_{\text{exp}} \sim H^{-1} \propto \frac{1}{g_*^{1/2}T^2}.$$

The freeze-out temperature of the n/p ratio is determined by equating this to the weak interaction timescale which depends on T^5. Thus freeze-out occurs earlier (at a higher temperature) for a larger g_* and the n/p ratio and hence the helium abundance is significantly enhanced. This formally limits, in particular, the number of types of relativistic neutrinos at the time of nucleosynthesis to less than 3.20 (Burles *et al* 1999). Experiments at CERN on the decay lifetime of the Z^0 determine the number to be three.

7.10   The plasma era

What are the interactions that couple the radiation field and matter together and for what range of redshift or temperature do they operate effectively?

7.10.1   Thomson scattering

The simplest process that is taking place between thermalization and the time when matter and radiation cease to interact is the low-energy scattering of photons by electrons. This process is Thomson scattering. The Thomson scattering timescale t_T is defined in terms of the mean free path λ_T by $\lambda_T = c t_T$. Thus

$$t_T = \frac{1}{\sigma_T n_e c},$$

where $\sigma_T = 6.65 \times 10^{-29}$ m^2 is the Thomson cross section for electrons.

For the moment we assume that the matter is fully ionized and composed entirely of electrons and protons for simplicity (so we ignore nuclei heavier than hydrogen). Then $n_e = n_e(t_0) T^3 / T_0^3$ with $n_e(t_0) = \rho_B / m_p = 1.88 \times 10^{-26} h^2 \Omega_B / m_p$, so

$$t_T = 0.9 \times 10^8 \left(\frac{T}{10^4 \text{ K}} \right)^{-3} (\Omega_B h^2)^{-1} \text{ s}. \tag{7.15}$$

This is to be compared with the expansion timescale. If we are in the matter-dominated phase, i.e. if $t > t_{eq}$, then t_{exp} is given by the second of (7.7). We obtain

$$\frac{t_T}{t_{exp}} \simeq (\Omega_B h^2)^{-1/2} \left(\frac{T}{20 \text{ K}} \right)^{-3/2}.$$

So $t_T / t_{exp} < 1$ as long as $T \gtrsim 20 (\Omega_B h^2)^{-1/3}$ K.

This is obviously consistent with our assumption of matter domination. (Had it not been we should have had to repeat the calculation in the radiation era to obtain a consistent solution.) Our assumption that the matter is fully ionized is, however, not valid at these temperatures, so the conclusion is that Thompson scattering couples the radiation and matter until ions and electrons recombine to neutral atoms. We shall investigate later when this occurs.

Note that the radiation interacts with the electrons, not the protons. This is because the Thomson scattering cross section for photon–proton scattering is a factor $(m_e/m_p)^2$ smaller, hence negligible. The effect of the radiation field is communicated to protons (and nuclei in general) by scattering with electrons to which they are strongly coupled by the usual Coulomb interaction.

7.10.2 Free–free absorption

The coupling of matter and radiation is complicated by the fact that we should not ignore absorption processes entirely. So called free–free absorption occurs when an electron absorbs a photon in the presence of a nucleus. The free–free absorption timescale is given by (Padmanabhan 1993)

$$t_{\text{ff}} = 3 \times 10^{14} (\Omega_B h^2)^{-2} \left(\frac{T}{10^4 \text{ K}} \right)^{-5/2} \text{ s} \tag{7.16}$$

for photons at the peak of the Planck spectrum having frequency $\nu \sim kT/h$. Equivalently the mean free path for free–free absorption is $\lambda_{\text{ff}} = ct_{\text{ff}}$. Comparing this with the expansion timescale in the matter-dominated era appears to show that free–free absorption is ineffective below 10^4 eV. This is not the case because of the effect of Thomson scattering. Consider a photon travelling from a point A to a point B. Suppose it has a small probability of absorption if it travels directly from A to B. However, if scattering forces it to travel in a zig-zag path it will spend longer in flight and suffer a higher probability of absorption.

We can estimate the extent of the effect by approximating the path as a one-dimensional random walk (although, in fact, it is obviously three-dimensional and the probabilities of scattering into all angles are not equal). Let λ_T be the mean free path of a photon to Thomson scattering and let λ_{ff} be the mean free path for free–free absorption. It is a basic property of random walks that after a number of steps, N, of a random walk of constant step length λ_T between scatterings a photon will have travelled a distance $N^{1/2}\lambda_T$. This is, therefore, the mean distance of travel we should expect after a large number of scatterings. The number of steps required to travel a distance λ_{ff}, which is the mean distance between absorptions, is $N = \lambda_{\text{ff}}/\lambda_T$. After this many scatterings the average photon has travelled a distance from its starting point $\bar{\lambda} = N^{1/2}\lambda_T = (\lambda_{\text{ff}}\lambda_T)^{1/2}$ in a time $N\lambda_T/c$, hence at an effective speed $(\lambda_T/\lambda_{\text{ff}})^{1/2}c$. Consider a thickness of material d. This has optical depth

$$\tau_{\text{abs}} = \frac{d}{\bar{\lambda}} = \left(\frac{d}{\lambda_{\text{ff}}} \right)^{1/2} \left(\frac{d}{\lambda_T} \right)^{1/2} = (\tau_{\text{ff}}\tau_T)^{1/2}.$$

The absorption timescale is the time such that $d = ct$ gives $\tau_{\text{abs}} = 1$, so

$$t_{\text{abs}} = \frac{\bar{\lambda}}{c} = (t_{\text{ff}}t_T)^{1/2}.$$

Using (7.15) for the scattering timescale and (7.16) for t_{ff} in a fully ionized medium gives

$$t_{abs} = 1.6 \times 10^{11} (\Omega_B h^2)^{-3/2} \left(\frac{T}{10^4 \text{ K}} \right)^{-11/4} \text{ s}$$

and

$$\frac{t_{abs}}{t_{exp}} = (\Omega_B h^2)^{-1} \left(\frac{T}{1900 \text{ K}} \right)^{-5/4},$$

so absorption appears to be important down to temperatures of order 1900 K or about 0.16 eV. In fact, it is not consistent to treat the gas as fully ionized down to these temperatures, and absorption ceases to be important once the plasma recombines at a temperature of about 4000 K.

7.10.3   Compton scattering

In fact, for most of the range of temperature down to 90 eV an alternative energy exchange mechanism dominates over absorption. This is Compton scattering in which a photon scattering off of an electron of speed v undergoes a change in frequency $\delta v / v \sim (v/c)^2$. Thomson scattering is the same physical process but with this effect, of order $(v/c)^2$, ignored.

In a thermal distribution of electrons with speeds v the average of v^2 is of order kT/m_e. Each scattering by an electron shifts the photon frequency by on average

$$\frac{\delta v}{v} \sim \frac{kT}{m_e c^2}.$$

The number of scatterings undergone by a photon which random walks through a distance ct_C is given by $N^{1/2} \lambda_T = ct_C$. Here λ_T is again the Thomson scattering mean free path since the process is governed by the same Thomson cross section. The photon therefore performs a random walk in frequency space, with step length $\delta v / v \sim kT/m_e c^2$. After N steps we expect to find the photon at a frequency shifted by Δv given by

$$\frac{\Delta v}{v} = N^{1/2} \left(\frac{kT}{m_e c^2} \right) = \left(\frac{ct_C}{\lambda_T} \right) \left(\frac{kT}{m_e c^2} \right).$$

The process is certainly important on a timescale t_C for which $\Delta v / v \sim 1$. This gives the Compton timescale

$$t_C = \frac{1}{\sigma_T n_e c} \left(\frac{m_e c^2}{kT} \right).$$

Comparing this with the Thomson scattering timescale we see that, in effect, the cross section has been reduced by a factor $kT/m_e c^2$. Finally, we have

$$\frac{t_C}{t_{exp}} \simeq (\Omega_B h^2)^{-1/2} \left(\frac{kT}{4 \text{ eV}} \right)^{-5/2}$$

so the process is important for maintaining equilibrium down to $T \sim 4$ eV or about 50 000 K.

7.11 Decoupling

As long as there are free electrons in sufficient abundance Compton scattering will keep the matter at the radiation temperature down to temperatures below 3000 K. If no other process were to intervene this would be the point below which the temperature of radiation and matter become entirely decoupled. In fact another process does intervene, namely the ions and electrons combine to form neutral atoms. This is known as recombination and is treated later. The neutral component of the matter is decoupled from the radiation and cools independently. On the other hand, the relatively few remaining free electrons are still coupled to the radiation and, although this has little effect on the radiation field, it can keep the electrons heated to the radiation temperature down to 60 K. To see this, note that the optical depths tell us what happens to the average photon but not how the average electron behaves. The relatively few electrons can and do make many collisions with photons, even though most of the photons do not collide at all. The relevant quantity is the mean free time of the *electrons* to Compton scattering, which is $(\sigma_T n_\gamma c)^{-1}$. The timescale for electrons to gain significant energy from photons by Compton scattering is therefore

$$t_{Ce} = \left(\frac{n_e}{n_\gamma}\right) t_C.$$

Taking the free electron density to be of order $10^{-5} n_B$, as estimated in the following section, it can be shown that

$$\frac{t_{Ce}}{t_{exp}} = (\Omega_B h^2)^{1/2} \left(\frac{T}{60 \text{ K}}\right)^{-5/2}.$$

Because $n_e \ll n_\gamma$ we have $t_{Ce} \ll t_{exp}$ right down to 60 K. Of course, the net energy transfer is zero once the electrons and photons are at the same temperature, so we conclude that Compton scattering keeps the electron component at the radiation temperature down to redshifts of 20.

7.12 Recombination

As the plasma cools electrons and nuclei start to recombine to form neutral atoms. If the plasma is in thermal equilibrium, with matter and radiation at the same temperature, or if the ionization equilibrium is maintained by balancing collisional ionizations with recombinations so that photoionization by the radiation field can be neglected, then the equilibrium degree of ionization is given by the Saha equation (appendix B). In fact, at recombination, neither of these conditions is a particularly good approximation. The matter and radiation are starting to decouple and hence to cool at different rates, and photoionization plays an important role in maintaining the ionization equilibrium, so a detailed kinetic treatment is required for an accurate description. Nevertheless we can use the Saha equation to give an order of magnitude estimate for the recombination epoch.

We consider only hydrogen since this is the dominant species. Let $x = n_e/n_B$ be the ratio of electron number density to the total baryon density at temperature T, and let $I_H = 13.6$ eV be the ionization energy of the H-atom. We have $n_p = n_e$ and $n_H = n_B(1 - x)$ and the Saha equation gives

$$\frac{1-x}{x^2} = 3.8\eta \left(\frac{kT}{m_e c^2}\right)^{3/2} \exp\left(\frac{I_H}{kT}\right). \tag{7.17}$$

where $\eta = 2.7 \times 10^{-8}(\Omega_B h^2)$ is the entropy parameter. We consider the plasma to be substantially recombined when $x = 0.1$ at which point 90% of the hydrogen atoms are neutral. Let the corresponding temperature be T_{rec}. Inserting numerical values gives

$$90 = 2.2 \times 10^{-22} T_{rec}^{3/2} \Omega_B h^2 \exp\left(\frac{1.6 \times 10^5}{T_{rec}}\right). \tag{7.18}$$

As usual, the factor $T_{rec}^{3/2}$ shifts T_{rec} away from the crude estimate $kT_{rec} \sim I_H$ to somewhat lower values, but the more substantial influence in the same downward direction comes from the large photon density through the small value of the factor η. The result is that, depending somewhat on the value of $\Omega_B h^2$, recombination occurs at around $T_{rec} \simeq 4000$ K or $z_{rec} \simeq 1300$. This can be obtained by iterative solution of (7.18). At lower redshifts we can generally consider the matter to be neutral.

Of course, as always we should check that the equilibrium can be achieved despite the expansion of the Universe. This we do by comparing the recombination timescale with the expansion timescale. The rate at which electrons of velocity v recombine with protons is given by

$$t_{rec}^{-1} \sim \sigma_{rec} v n_e$$

per proton, where $\sigma_{rec}(v)$ is the recombination coefficient for an electron of velocity v. Averages of the product $\langle \sigma_{rec} v \rangle$ for a thermal plasma are tabulated as recombination coefficients $\alpha(T)$ (e.g. Allen 1973). With $\alpha(T) \sim 10^{-16} T^{-1/2} \, m^3 \, s^{-1}$ we can calculate the mean recombination time $t_{rec} = (\alpha n_e)^{-1}$ at T_{rec} to confirm that $t_{rec} < t_{exp}$ at recombination. Note however that this argument depends on the value of n_e at recombination. We can turn this argument round to find the density of free electrons below which recombination cannot occur on less than an expansion timescale. This will occur at the temperature for which $t_{rec} = t_{exp}$. The corresponding equilibrium electron fraction turns out to be around 10^{-5}. This is the minimum residual ionization of the intergalactic plasma.

7.13   Last scattering

We saw earlier that, to order of magnitude, photons and electrons are strongly coupled as long as the Thomson scattering time is less than the expansion time, or, equivalently, up to the time t given by $\sigma_T n_e c t = 1$ when the optical depth to the Hubble sphere reaches unity. To investigate what we see in looking back to that time we consider that photons will reach us typically from an optical depth of unity along a ray. This defines the time of last scattering t_{ls} from

$$\int_0^{t_{ls}} \sigma_T n_e c \, dt = 1. \tag{7.19}$$

To evaluate the integral we need the evolution of the electron density n_e as a function of time, or redshift. This is not a simple matter since we need to follow the time dependence of the recombination process. It will be further complicated if processes have occurred since recombination that have re-ionized the intergalactic medium, for example the UV emission of quasars soon after their formation. The time of last scattering can be estimated from an approximate evaluation of the integral in (7.19). The outcome of detailed calculations (see e.g. Padmanabhan 1993) is that as we look back the optical depth rises sharply at recombination at redshifts of about 1100, and that most of the photons we receive were last scattered in a narrow range around this.

7.14   Perturbations

The recombination of ions and electrons produces photons which add to the background radiation but cannot be thermalized. In principle, this effect should be observable as a distortion of the Planck spectrum. If we consider only recombinations directly to the ground state then each recombination releases an energy $I_H \sim 13.6$ eV and there are n_e recombinations per unit volume at redshift z_{rec}. By integrating the Planck spectrum (problem 66) we find about 7000 photons m^{-3} above 13.6 eV compared with $n_e(1 + z_{rec})^3 \sim 10^{10}\Omega_B h^2$ photons m^{-3} from recombination. Unfortunately this significant distortion of the blackbody spectrum occurs in the far infrared where it is swamped by emission from galactic dust.

At the other extreme we can consider the production of nuclear energy during nucleosynthesis as a potential addition to the background heat. In the worst possible case we can assume that all the fusion energy goes into radiation. The energy per unit volume released as a fraction of the background energy density is given by

$$\frac{\text{number density of } {}^4\text{He} \times \text{binding energy of } {}^4\text{He}}{\text{energy density of photons}}$$

$$= \frac{0.06\eta E_B}{2.7kT} = 1.8 \times 10^{-7}\Omega_B h^2.$$

This is obviously negligible and will, in any case, be thermalized. On the other hand, limits can be set on hypothetical particle decays from the lack of distortions of the background radiation.

Finally, a non-uniform distribution of hot gas in the line of sight between us and decoupling can Compton scatter photons to higher energies and distort the spectrum. This is known as the Zeldovich–Sunyaev effect. Precisely such small distortions are seen where rich clusters of galaxies lie in the line of sight, the effect being caused by the hot intracluster gas.

7.15   Appendix A. Thermal distributions

For a perfect gas in thermal equilibrium the number density $n_i(p)\,\mathrm{d}p$ of particles of type i with momentum between p and $p + \mathrm{d}p$ is

$$n_i(p)\,\mathrm{d}p = g_i \frac{4\pi p^2 \,\mathrm{d}p}{h^3} \frac{1}{\left[\exp\left(\frac{E_i(p)-\mu_i}{kT}\right) \pm 1\right]} \qquad (7.20)$$

where μ_i is the chemical potential of species i, related to the overall number density of that species and g_i the statistical weight (or number of internal degrees of freedom). With the minus sign in the denominator the formula applies to bosons, and with the plus sign to fermions. The quantity $E_i(p)$ is the energy per particle of species i as a function of momentum, so for a particle with rest mass m_i

$$E_i(p) = (c^2 p^2 + m_i^2 c^4)^{1/2}.$$

7.15.1   Chemical potentials

In the text we have neglected certain chemical potentials of particle species in the early Universe. In this section we consider the justification for this.

Note first that the chemical potential of a species is the free energy per particle and that the free energy of a system in thermal equilibrium is a minimum. Thus, for a reaction of the form

$$A + B \rightleftharpoons C + D$$

we have, at equilibrium,

$$\mu_A + \mu_B = \mu_C + \mu_D$$

since otherwise we could reduce the overall free energy by interconverting more particles.

On the other hand, if a particle species is not conserved its free energy must be zero. Formally we can see this by noting that the free energy is stationary at constant temperature and volume if

$$0 = \left[\frac{\partial F}{\partial N}\right]_{T,V} = \mu.$$

Hence, for example, $\mu_\gamma = 0$ if photons can be freely created.

We can now deduce from the reaction $e^+ + e^- \rightleftharpoons \gamma + \gamma$ that $\mu_{e^+} + \mu_{e^-} = 0$ and from $e^+ + e^- \rightleftharpoons \nu_e + \bar{\nu}_e$ that $\mu_\nu + \mu_{\bar{\nu}} = 0$. Therefore the chemical potential of a particle has the same magnitude but opposite sign from that of its partner antiparticle.

If the numbers of neutrinos and antineutrinos in the early Universe are equal then from equation (7.20) and using $\mu_\nu = -\mu_{\bar{\nu}}$, we have

$$n_\nu - n_{\bar{\nu}} = \frac{g}{h^3} \int 4\pi p^2 \, dp \left\{ \left[\exp\left(\frac{pc - \mu_\nu}{kT}\right) + 1 \right]^{-1} \right.$$
$$\left. - \left[\exp\left(\frac{pc + \mu_\nu}{kT}\right) + 1 \right]^{-1} \right\} = 0.$$

This equation can be satisfied only if $\mu_\nu = \mu_{\bar{\nu}} = 0$.

As we cannot detect the neutrino background we cannot establish directly whether there are equal numbers of neutrinos and antineutrinos for all three neutrino families. But it seems plausible and is a basic assumption of the standard big-bang cosmology. The best test is that calculations of element abundances using this assumption appear to be in accord with observation.

When the electron–positron pairs are in thermal equilibrium at $T > 10^{10}$ K there is a small excess of electrons over positrons. These extra electrons eventually pair with protons and provide overall charge neutrality. Therefore we have

$$n_{e^-} - n_{e^+} = n_B$$

where $n_B/n_e \sim \eta$ is a small quantity.

Consider now the ratio of neutrons to protons in the early Universe. At temperature $T \sim 10^{10}$ K these form a non-relativistic classical gas, so we can obtain the number densities from (7.20) in the limit of small occupancy. This condition requires

$$\exp\left(\frac{E - \mu}{kT}\right) + 1 \gg 1,$$

for which (7.20) becomes

$$n_i(p) \, dp = \frac{g_i}{h} 4\pi p^2 \, dp \, e^{\mu/kT} e^{-E/kT},$$

where $E = mc^2 + p^2/2m$ for a non-relativistic gas with $p \ll mc$. Integrating over E to find the density of species i gives

$$n_i = g_i \left(\frac{mkT}{2\pi}\right)^{3/2} e^{(\mu - mc^2)/kT}.$$

So the thermal equilibrium ratio of neutrons to protons is

$$\frac{n_n}{n_p} = \left(\frac{m_n}{m_p}\right)^{3/2} \exp\left(\frac{\mu_n - \mu_p - m_n c^2 + m_p c^2}{kT}\right).$$

Now the processes that maintain equilibrium between protons and neutrons before the freeze-out of the weak interaction are

$$\nu_e + n \rightleftharpoons e^- + p$$
$$\bar{\nu}_e + p \rightleftharpoons e^+ + n.$$

From the first reaction the chemical potentials satisfy

$$\mu_\nu + \mu_n = \mu_{e^-} + \mu_p.$$

Hence

$$\frac{n_n}{n_p} = \left(\frac{m_n}{m_p}\right)^{3/2} \exp\left(\frac{\mu_{e^-} - \mu_\nu - Q}{kT}\right), \qquad (7.21)$$

where $Q = (m_n - m_p)c^2$.

Now, as we explained earlier, the chemical potential of the neutrinos is taken to be zero. Note, however, that a non-zero value would affect the neutron-to-proton ratio, making it smaller or larger depending on the sign of μ_{e^-}. This would shift the amount of helium produced away from the value of 25% by mass we obtained in section 7.9.2. If the chemical potentials of the other neutrinos were non-zero this would affect the mass-energy density, giving a value exceeding $7aT^4/8$. This would reduce the expansion timescale through a larger value of g_* changing the frozen in value of n_n/n_p.

Finally, we can show that $\mu_{e^-}/kT \sim \eta$, which is much less than Q/kT (problem 62). We can, therefore, drop this term from the exponent of (7.21) leaving us with the expression used in the text.

7.15.2   Photon energy density

For photons (hence for blackbody radiation), $E_\gamma(p) = cp = h\nu$, $\mu_\gamma = 0$, and $g_\gamma = 2$, so

$$n(E)\,dE = \frac{8\pi}{h^3}\frac{E^2\,dE}{c^3}\frac{1}{\exp\left(\frac{E}{kT}\right) - 1}.$$

The energy density is

$$u(E)\,dE = En(E)\,dE,$$

so the overall energy per unit volume is given by

$$u_\gamma = \frac{8\pi}{c^3 h^3}\int_0^\infty \frac{E^3\,dE}{\exp\left(\frac{E}{kT}\right) - 1}$$

$$= \frac{8\pi}{c^3 h^3}k^4 T^4 \int_0^\infty \frac{x^3\,dx}{e^x - 1},$$

on putting $E = xkT$. The integral is a numerical constant ($\pi^4/15$ in fact, although this cannot be obtained by elementary methods) so we get

$$u_\gamma = aT^4,$$

where $a = 8\pi^5 k^4/(15h^3c^3) = 7.56 \times 10^{-16}$ J m^{-3} K^{-4} is the radiation constant.

7.15.3 Photon number density

The number density of blackbody photons is given by

$$n_\gamma = \frac{8\pi}{c^3h^3} \int_0^\infty \frac{E^2 \, \mathrm{d}E}{e^{\frac{E}{kT}} - 1},$$

$$= \frac{8\pi}{c^3h^3} k^3 T^3 \int_0^\infty \frac{x^2 \, \mathrm{d}x}{e^x - 1}.$$

The integral is a numerical constant which can be evaluated (again not by elementary methods) as $2 \times \zeta(3) \simeq 2.4$, where ζ is the Riemann zeta-function. This gives

$$n_\gamma = \frac{aT^4}{2.7kT} = \frac{aT^3}{2.7k}.$$

7.15.4 Relativistic neutrinos

At high temperatures we can assume that neutrinos are massless. Then $\nu\bar{\nu}$ pairs can be created freely in thermal equilibrium, so the number of neutrino pairs is not conserved. Thus we have $\mu_\nu + \mu_{\bar{\nu}} = 0$. If, furthermore, the numbers of neutrinos and antineutrinos are the same, then $\mu_\nu = \mu_{\bar{\nu}} = 0$.

Next $g_\nu = 1$ because neutrinos exist only in left-handed polarization states. Unlike the photon, neutrinos and antineutrinos are different particles and the number densities and energy densities for neutrino species are usually given for particle–antiparticle pairs. The energy of a zero mass particle is $E_\nu = cp$, so the overall energy density in thermal equilibrium at temperature T becomes

$$u_{\nu\bar{\nu}} = \frac{8\pi}{c^3h^3} k^4 T^4 \int_0^\infty \frac{x^3 \, \mathrm{d}x}{e^x + 1}.$$

The numerical value of the integral can be obtained by the following trick: we have

$$\int_0^\infty \left(\frac{x^3}{e^x - 1} - \frac{x^3}{e^x + 1} \right) \mathrm{d}x = \int_0^\infty \frac{2x^3}{e^{2x} - 1} \, \mathrm{d}x$$

$$= \frac{1}{2^3} \int_0^\infty \frac{z^3}{e^z - 1} \, \mathrm{d}z,$$

where the final integral is obtained by changing the variable to $z = 2x$. Rearranging gives

$$\int_0^\infty \frac{x^3}{e^x + 1} \, \mathrm{d}x = \left(1 - \frac{1}{2^3} \right) \int_0^\infty \frac{x^3}{e^x - 1} \, \mathrm{d}x = \frac{7}{8} \frac{\pi^4}{15}.$$

Finally therefore, for each neutrino–antineutrino species

$$u_{\nu\bar{\nu}} = \tfrac{7}{8}aT^4.$$

Similarly we can show

$$\int_0^\infty \frac{x^2}{e^x+1}\,dx = \left(1 - \frac{1}{2^2}\right)\int_0^\infty \frac{x^2}{e^x-1}\,dx = \frac{3}{4} \times 2\zeta(3),$$

where $\zeta(3) \simeq 1.2$ is the Riemann zeta-function, from which we can obtain the neutrino–antineutrino number density

$$n_{\nu\bar{\nu}} = \frac{aT^3}{3.6k}.$$

For massive neutrinos these results apply at high temperatures where we can neglect the rest mass-energy (called the ultrarelativistic limit).

7.15.5   Relativistic electrons

At high temperatures relativistic electron–positron pairs can be created freely in equilibrium with the radiation field so $\mu_e = \mu_{e^+} = 0$. For spin-$\tfrac{1}{2}$ particles there are two polarization states, which, for electrons and positrons, are both possible. Therefore $g_e = g_{e^+} = 2$. In the ultrarelativistic limit we neglect the rest masses so, also, $E_{e^\pm} = cp$. The energy densities and number densities in this limit are similar to those for neutrinos, except for the factor 2 from the statisitical weights. Therefore

$$u_e = u_{e^+} = \tfrac{7}{8}aT^4$$

and

$$n_e = n_{e^+} = \frac{aT^3}{3.6k}.$$

7.15.6   Entropy densities

The entropy density of a particle species in thermal equilibrium is found from

$$ds = \frac{du}{T},$$

which is equivalent to $s = (4/3)u/T$ if $u \propto T^4$.

This gives the following:

$$s_\gamma = \tfrac{4}{3}aT_\gamma^3, \qquad s_{e^\pm} = \tfrac{4}{3} \times \tfrac{7}{4}aT^4, \qquad s_{\nu\bar{\nu}} = \tfrac{4}{3} \times \tfrac{7}{8}aT^4. \tag{7.22}$$

7.16   Appendix B. The Saha equation

The Saha equation gives the fraction of ionized atoms as a function of electron density and temperature. For a pure hydrogen plasma in thermal equilibrium at temperature T, let n_H be the number of atoms of atomic hydrogen, and n_p and n_e be the numbers of free protons and electrons. Then the Saha equation gives

$$\frac{n_p n_e}{n_H} = \left(\frac{m_e kT}{2\pi}\right)^{3/2} \exp\left(-\frac{I_H}{kT}\right)$$

where $I_H = 13.6$ eV is the ionization energy of the H-atom. Writing $x = n_e/n_B$ for the ratio of electron number density to the total baryon density, then since $n_p = n_e$ we have also $n_H = n_B(1-x)$ and the Saha equation can be written

$$\frac{x^2}{1-x} = n_B^{-1}\left(\frac{2\pi m_e kT}{h^2}\right)^{3/2} \exp\left(-\frac{I_H}{kT}\right)$$

or, in terms of the entropy parameter $\eta = 2.7 \times 10^{-8}(\Omega_B h^2)$,

$$\frac{1-x}{x^2} = 3.8\eta \left(\frac{kT}{m_e c^2}\right)^{3/2} \exp\left(\frac{I_H}{kT}\right). \tag{7.23}$$

7.17   Appendix C. Constancy of η

Strictly speaking η has been constant only since the annihilation of electron–positron pairs which was completed somewhat below 10^{10} K. A more general treatment uses the total entropy inherited from early times, which, at present, resides in the three families of neutrinos as well as in the photons. Thus

$$s_{tot} = s_\gamma + s_{\nu\bar{\nu}}.$$

The total entropy is believed not to have changed appreciably since it was created in the very early Universe. Similarly, the baryon number is a relic of very early times. So we can regard the total entropy per baryon s_{tot}/kn_B as a parameter of the big-bang Universe. An aim of the cosmology of the very early Universe is to understand why this quantity has its observed value (chapter 9).

We now calculate s_{tot}/kn_B in terms of quantities at the present time. For this we need the current value of the energy density of each neutrino family, which is

$$u_{\nu\bar{\nu}} = \tfrac{7}{8}aT_\nu^4,$$

for a neutrino temperature T_ν (see appendix A). From section 7.7.3 we derived $T_\nu = (4/11)^{1/3}T_0$ and from section 7.15.6 $s_{\nu\bar{\nu}} = (4/3)(u_{\nu\bar{\nu}}/T_\nu)$ so the entropy of the three neutrino families is

$$s_{\nu\bar{\nu}} = 3 \times \frac{4}{3}\frac{u_{\nu\bar{\nu}}}{T_\nu} = \frac{4}{3} \times \frac{84}{88}aT_0^3.$$

So, finally,

$$\eta_{\text{tot}}^{-1} = \left(\frac{s_{\text{tot}}}{kn_b}\right)_0 = \frac{s_\gamma + s_{\nu\bar{\nu}}}{kn_B} = \frac{86}{33}\frac{aT_0^3}{kn_B}$$

$$= 2.57 \times 10^8 (\Omega_B h^2)^{-1}.$$

The constancy of this quantity provides a useful relationship for n_b as a function of radiation temperature T (see problem 60):

$$n_b = \frac{2\eta_{\text{tot}}g_* aT^3}{3k}$$

$$= 0.14g_* T^3 \Omega_B h^2.$$

7.18 Problems

Problem 53. *By considering the ratio of electrostatic potential energy to kinetic energy, show that the ideal gas approximation holds in the early Universe despite the high densities.*

Problem 54. *Assume that the primordial material before the start of stellar nucleosynthesis was hydrogen. Taking the luminosity density due to galaxies to be a constant $2 \times 10^8 L_\odot$ Mpc^{-3}, estimate the mass fraction of this hydrogen that would be converted into helium in stars by the present time.*

Problem 55. *The mean free path of a photon in the present Universe depends on the density and state of ionization of the intergalactic medium. Estimate a lower bound to the mean free path by taking all the baryonic matter in the present Universe to be uniformly distributed and in a fully ionized form. (The Thomson cross section is $\sigma_T = 6.6 \times 10^{-29}$ m^2.)*

Problem 56. *A fraction r of X particles decay to two quarks (qq) with baryon number 2/3, while the remaining fraction $1 - r$ decay to an antiquark and antilepton ($\bar{q}\bar{l}$) with baryon number $-1/3$. Similarly, $\bar{X}$ particles decay to $\bar{q}\bar{q}$ (baryon number $-2/3$) with branching ratio $\bar{r}$ and to ql (baryon number 1/3) with branching ratio $1 - \bar{r}$. The baryon number produced by the decay of an X is therefore $(2/3)r + (-1/3)(1 - r) = r - 1/3$. Calculate the baryon number produced by the decay of an $\bar{X}$ and hence show that the net baryon production per $X\bar{X}$ decay is $r - \bar{r}$. If X and $\bar{X}$ are produced in thermal equilibrium and decay when $kT \ll m_X c^2$ show that this produces an entropy per baryon ratio of $g_* k/(r - \bar{r})$ (see Kolb and Turner 1990, p 161).*

Problem 57. *Show that the energy density at the present time of three massless families of neutrinos is $0.68aT_0^4$ (where T_0 is the current radiation temperature).*

Problem 58. *Show that the redshift of matter radiation equality is given by* $1 + z_{eq} = 2.4 \times 10^4 \Omega_M h^2$.

Problem 59. *(a) The exact expression for t_{eq} quoted in section 7.4 can be obtained by integrating the Friedmann equation with $\rho = \rho_r + \rho_m$. To do this write $\rho_r = \rho_{eq} R_{eq}^4 / R^4$, and $\rho_m = \rho_{eq} R_{eq}^3 / R^3$, separate the variables and integrate from $t = 0$ to $t = t_{eq}$. Any contribution from the cosmological constant or the curvature term will be negligible at these epochs. Show that this gives*

$$t_{eq} = \frac{4(\sqrt{2} - 1)}{3 H_{eq}}.$$

(b) Starting from $\Omega_m + \Omega_r = 1$, show that

$$H_{eq} = H_0 [2(1 + z_{eq})^3 \Omega_M]^{1/2}.$$

(c) Finally show that

$$t_{eq} = 1.03 \times 10^3 (\Omega_M h^2)^{-2} \text{ years}.$$

Problem 60. *Show that the density in baryons at the epoch when the temperature is T (in Kelvin) is*

$$n_b = 0.14 g_* T^3 \Omega_B h^2.$$

(One way to do this problem is to use conservation of entropy.)

Problem 61. *The neutron-to-proton ratio n/p at the time of neutron freeze-out ($t_{np} \sim 1$ s) is 1/6. Up to the time of nucleosynthesis t_n neutron β-decay reduces this ratio. Show that at the time of helium production at $kT \sim 0.1$ MeV ($t \sim 3$ min) the n/p ratio is reduced to about 1/7 and hence calculate the corrected helium abundance (see Padmanabhan 1993, p 105).*

Problem 62. *Starting from the difference in densities of electrons and positrons*

$$n_{e^-} - n_{e^+} = \frac{8\pi}{h^3} \int_0^\infty p^2 \, dp \left[\frac{1}{\exp[(pc - \mu_e)/kT] + 1} - \frac{1}{\exp[(pc + \mu_e)/kT] + 1} \right],$$

assuming this to be small, and expanding in the small quantity μ_e/kT, show that the entropy parameter is

$$\eta \simeq \frac{\mu_e}{kT}.$$

Note that

$$\int_0^\infty \frac{x^2 e^x}{(1 + e^x)^2} \, dx = \int_0^\infty \frac{x^2 e^{-x}}{(1 + e^{-x})^2} \, dx \simeq \int_0^\infty x^2 e^{-x} \, dx \simeq 2.$$

Problem 63. *Massive neutrinos which are relativistic at the time that the weak interactions freeze out will have a number density at the present time equal to that of massless neutrinos,*

$$n_{\nu\bar{\nu}} = \tfrac{3}{11} n_\gamma,$$

and will contribute a mass density $\rho_{\nu\bar{\nu}} = m_\nu n_{\nu\bar{\nu}}$. If one neutrino family contributes the critical density, compute the mass of this neutrino type. (The masses of the neutrinos are not known, but non-zero values of less than 1 eV seem to be suggested by present evidence, insufficient to provide the critical density.)

Problem 64. *Show that the present entropy density in photons and neutrino families (assumed massless) is*

$$s_{\rm T} = \left(\tfrac{43}{22}\right) \tfrac{4}{3} a T_0^3.$$

Before $e^+ - e^-$ annihilation the entropy density was $\tfrac{4}{3} g_* a T^3$. Assuming entropy is conserved in a comoving volume, show that $RT = (43/11g_*)^{1/3} R_0 T_0$ (i.e. the RT product is not conserved at early times).

Problem 65. *(a) For photons, neutrinos and electrons (and their corresponding antiparticles) in equilibrium at temperature T, show that the energy density is*

$$u_{\rm T} = \frac{g_*}{2} a T^4,$$

where $g_* = 43/4$, and that the entropy density is $s_{\rm T} = \tfrac{4}{3}(u_{\rm T}/T)$.
   (b) For a mixture of photons at temperature T and neutrinos at temperature T_ν, find the corresponding energy and entropy densities and show that $s_{\rm T} \neq \tfrac{4}{3}(u_{\rm T}/T)$.

Problem 66. *Show that the number density of photons in the blackbody background at recombination (around 3000 K) with energies above 13.6 eV is about 7000 m^{-3}.*

Problem 67. *Show that the electron density at recombination is approximately $10^{10}\Omega_B h^2 \ m^{-3}$. Show that recombination photons at 13.6 eV are shifted to a wavelength of 100 μm in the infrared at the present time.*

Chapter 8

Inflation

The problems of the standard model that we outlined at the end of chapter 6 have a notable similarity. They all require for their resolution a period of exceptionally rapid expansion or, equivalently, a dilution of the contents of the Universe. Interposing such an expansion or dilution between ourselves and the beginning of the Universe will reduce the curvature (the flatness and age problems), and expand the region of space that can arise from a causally connected region (the horizon problem and the problem of structure formation). With the usual power law expansion ($R(t) \propto t^p$, $0 < p < 1$) the expansion is slowing down ($q > 0$). If $p > 1$, then the deceleration parameter is negative and the expansion is speeding up. By an exceptionally rapid expansion it is usually implied that the expansion must be speeding up by more than this, so the expansion is more rapid than a power law dependence on time, for example, an exponential expansion. So an inflationary model of the Universe is one with such a period of exceptionally rapid expansion.

By itself such a 'solution' creates two further problems. First, the Universe is not, at the present time, in an inflationary phase. Therefore, a viable theory must offer a way of bringing inflation to an end. Second, if the dilution were to apply to everything we should be left without the matter and radiation of the present Universe. So there must be some way of regenerating the matter content after sufficient inflation has occurred. In this chapter we shall show how inflation models achieve this.

By an inflationary expansion of about a factor 10^{30}, the problems of the big-bang theory can be ameliorated to such an extent that many cosmologists would regard them as solved. Inflation models can easily achieve this amount of expansion. This remarkable fact drives the theory on in the absence of any experimental evidence. One must be aware, however, that knowing what a theory can do for you, if true, does not make it true. Thus, cosmologists turn to the properties of matter under conditions appropriate to the early Universe, hoping to find there the components of inflation. We believe it fair to say that the guesses as to what these properties might be do not yet provide a complete and consistent theory of inflation, and it is not beyond the realms of possibility that the difficulties of the big-bang theory could be resolved in other (quantum) ways.

8.1   The horizon problem

The horizon problem arises because the integral for the proper distance to the horizon (equation (5.84))

$$D_h = cR(t) \int_T^t \frac{dt'}{R(t')}$$

converges as T tends towards early times. If, however, the Universe was accelerating in the past, so $\ddot{R} > 0$, then the integral diverges. It is easy to show this in the simplest case of exponentially increasing acceleration

$$R = R_0 e^{Ht}$$

for $-\infty < t < \infty$, where H is a constant, which corresponds to a constant deceleration parameter $q_0 = -1$. This is the steady-state model, but now interpreted as an inflationary Universe containing matter with an unusual equation of state (see later) rather than with the continuous creation of particles of normal matter. For this case

$$D_h = \lim_{T \to -\infty} R(t) \int_T^t \frac{dt'}{R(t')} = \lim_{T \to -\infty} \frac{e^{-HT} - e^{-Ht}}{H} \to \infty.$$

Problem 68 shows how this works if $\ddot{R}$ is positive but not constant.

In these examples the Universe is taken to be accelerating for all time. But the resolution of the horizon problem does not imply that we have to take the models seriously right back to the big bang at $t = 0$ (or even to $t = -\infty$) where the assumptions of classical physics do not apply. What it does imply is that, under the circumstances of accelerated expansion, the horizon distance can acquire a value *sufficiently large* that the observable Universe is a small part of what is a large, but finite, causally connected uniform region. In models with $\Lambda = 0$ now, this solves the horizon problem for a large, but finite time. If the inflationary picture is true then, at some very long time in the future, we shall be able to see to larger distances and the inhomogeneous edge of our causal patch will reveal itself. On the other hand, if $\Lambda > 0$, or in the presence of quintessence, the Hubble sphere will cease to overtake galaxies and we shall never see galaxies beyond it. The Universe will be in a permanent state of 'mild inflation'. (Unless, of course, there is a further change in the dark energy such that the Universe becomes matter dominated once again, as can happen in some quintessence schemes; see section 5.20.)

If, however, we can extrapolate back to times early enough that the Universe as a whole was a quantum system, a period of inflation is at least consistent with the speculation that our patch of the Universe began as a quantum fluctuation (Tryon 1973). By contrast, in non-inflationary models early quantum fluctuations cannot grow on sufficient scales.

8.2   The flatness problem

The flatness problem arises because the value $\Omega = 1$ is an unstable equilibrium. Any departure from $\Omega = 1$ at some time, however small, eventually grows to be large. As with the horizon problem, inflation puts off this growth to a long time into the future. To see this, we write the Friedmann equation (5.33) in the form

$$|\Omega - 1| = \frac{|k|c^2}{R^2 H^2}.$$

Then $\ddot{R} > 0$ implies

$$\frac{\mathrm{d}}{\mathrm{d}t}|\Omega - 1| = \frac{\mathrm{d}}{\mathrm{d}t}\left(\frac{|k|c^2}{R^2 H^2}\right) = \frac{\mathrm{d}}{\mathrm{d}t}\left(\frac{|k|c^2}{\dot{R}^2}\right) = -\frac{2|k|\ddot{R}c^2}{\dot{R}^3} < 0.$$

So as the Universe evolves through a period of inflation $|\Omega - 1|$ gets smaller, hence closer to zero, and hence Ω is driven towards unity, rather than further away. This convinced many cosmologists that there should be a contribution to Ω from dark matter which would give $\Omega \simeq 1$. (It also convinced some, incorrectly, that Ω should be exactly 1.) Note that, if the expansion were not accelerating at the present time, $|\Omega - 1|$ would now be departing again from zero and the flatness 'problem' would resurface again in the future.

8.3   Origin of structure

Despite the appeal of eliminating the horizon and flatness problems, one of the most attractive features of inflation is now taken to be the possibility of understanding how structure arises in the Universe. Recall that the difficulty in the big-bang theory is that the region of space occupied by, say, a galaxy now was larger than a causally connected region at sufficiently early times. For example, in the Einstein–de Sitter model an atom on the Hubble sphere at a distance $l_{\mathrm{h}} = 3ct/2$ is moving away from us at the speed of light (section 5.18). But a part of the Universe with a galactic mass (containing a galaxy now, but dispersed at earlier times) has a radius $R(t)l_0/R(0) \propto t^{2/3}$, and hence extends beyond its Hubble sphere at sufficiently early times. This implies that galaxy size perturbations could not have arisen from causal processes. In other words the structure of the Universe would have to have been programmed into the initial conditions. (If you believe that galaxies were formed by assembling smaller sub-units then the same argument applies to these.)

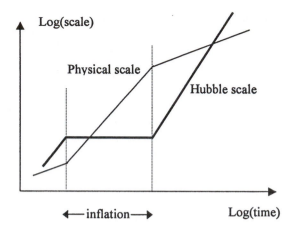

Figure 8.1. Schematic evolution of physical scales and the Hubble sphere.

Figure 8.1 shows the difference that inflation makes. Relevant physical scales now fit comfortably inside the Hubble sphere at early times and perturbations on these scales can be generated by causal physics. As the Universe inflates a physical scale increases by the ratio of scale factors, hence proportional approximately to e^{Ht}. On the other hand, the physical Hubble length, which encompasses objects with relative speeds less than that of light, is $c/H \sim$ constant. A physical scale, therefore, expands relative to the Hubble scale, and hence becomes larger than a causal region. The perturbations remain intact until after inflation has ended, when once again a physical scale starts to grow as $R(t) \propto t^{1/2}$, while the Hubble sphere now grows as $H^{-1} \propto t$. A physical scale therefore shrinks relative to the Hubble sphere so a perturbation will eventually come within the speed of light surface and, subject to pressure forces, evolve to form structure. Note that a perturbation on a smaller scale than the one shown in the figure crosses the Hubble sphere later during inflation but re-enters earlier once inflation has ended. This is as we would expect, since it takes longer to inflate a smaller scale to a given size, but smaller scales fit into a standard expanding Universe earlier.

Note that we refer throughout this discussion to the Hubble sphere, or, equivalently to the speed of light surface at a distance $d_{\mathrm{H}} = c/H$. Many authors speak instead of the horizon, meaning the Hubble sphere. Strictly the horizon is a different surface which divides us from galaxies with which we have not yet had causal contact. In a radiation-dominated Universe the horizon is at $2ct$, hence it is the same size as the Hubble sphere (see also Harrison 2000; Ellis and Rothman 1993).

During inflation the Hubble sphere is stationary, as we can see from the equation

$$\frac{\mathrm{d}d_{\mathrm{H}}}{\mathrm{d}t} = c(1+q),$$

with $q = -1$ (section 5.18). So scales of interest expand beyond the Hubble distance while inflation is going on. At the end of inflation, when normal expansion is resumed, we have $q > 0$ and the Hubble sphere expands faster that the matter again. These scales therefore come back within the Hubble sphere in due course and are subject to causal influence.

The favourite candidate for the origin of the perturbations that yield large-scale structure is quantum fluctuations. In this picture therefore galaxies are the most visible manifestation of the Heisenberg uncertainty principle. As we shall see in chapter 9, the challenge is to find an inflation model that produces just the right amplitude of fluctuation on each length scale to agree with the observed structure.

8.4 Mechanisms

Experiments cannot be extrapolated, only theories. Thus the investigation of the early Universe can proceed only so far as we trust our theories. We trust recent discoveries to this extent: we believe that the phases of matter at high densities and temperatures are qualitatively different from our low-density, low-temperature experience. To see the significance of this, imagine that we lived in the sea. We might be led to believe that matter existed in just two phases: liquid and solid. Only through physical theory could we predict the gaseous state: and we should have *no other concept of what that state would be like*. So we have to begin by understanding phases in general, and in the context of particle physics in particular.

The particular phase changes of interest are those associated with the breaking or restoration of symmetry. The most familiar example is that of a ferromagnet. At low temperatures the atomic magnets in a domain of a ferromagnetic material line up, despite the fact that the equations of electromagnetism do not specify a preferred direction. We say that the symmetry (of Maxwell's equations of electromagnetic theory which contain no special direction) is broken by the ground state. At temperatures above the Curie temperature, the symmetry is restored and there is no net magnetic moment. Note that the special direction of the ground state is arbitrary, so the symmetry is broken in any one instance, but not in an ensemble. This is shown in our example by the existence of ferromagnetic domains which align randomly (unless one supplies a preferred direction in the form of an external magnetic field).

There are two facts of particle physics that dominate the relevance of this for cosmology. The first is that a symmetry restored ground state can correspond to a state of high energy relative to the broken symmetry vacuum. At high temperature, when the symmetry is unbroken, this additional energy density affects the cosmological expansion. The energy of this state appears as latent heat as the Universe cools, and symmetry is broken, and can provide the radiation entropy we observe now. Note that during inflation the Universe evolves from a lowest energy state to a new lowest energy state, so from a (symmetrical) vacuum state to a (non-symmetrical) vacuum state. The energies of these states are different, but both contain no normal matter. Cosmologists often speak of a false vacuum and the true vacuum for the two states.

The second fact is that symmetry breaking may leave relics. These might be useful seeds for galaxy formation or they might be unobserved entities such as magnetic monopoles. We do not know in detail, because we do not know the symmetries of particle physics beyond the standard model. However, any model that incorporates symmetry breaking will suffer from this problem. The big idea, however, is to use the first fact to resolve any possible problems with the second. Specifically, the theory of inflation tries to arrange that the energy density in the symmetric ground state drives an exponential expansion of the Universe so as to dilute, essentially to zero, the density of any unwanted relics.

Having obtained the main idea of a large false vacuum energy from particle physics one can proceed to develop models of inflation that ignore the particle physics. This leads to various alternative models of inflation. We shall therefore proceed to a general development of inflation models and return to the particle physics in later sections.

8.4.1   Equation of motion for the inflaton field

If the vacuum energy that drives inflation is to arise from a phase transition it must involve spin-zero particles so as not to single out a direction in space. Classically such particles are represented by a scalar field, in much the same way that photons appear classically as the (vectorial) electromagnetic field. Consider such a scalar field ϕ, say, in a spatially homogeneous Universe, so $\phi = \phi(t)$. The field has a kinetic energy density $\frac{1}{2}\dot{\phi}^2$. It can also have a potential energy density which arises from the self-interaction of the field, and appears as an algebraic function $V(\phi)$. The total energy in a comoving volume R^3 is

$$E = (\tfrac{1}{2}\dot{\phi}^2 + V(\phi))R^3. \tag{8.1}$$

Think of ϕ as a particle coordinate x, so this is the energy of a particle with time-dependent mass R^3 in a time-varying potential $R^3 V(x)$. The particle momentum is $R^3\dot{x}$, so the equation of motion is

$$\frac{\mathrm{d}}{\mathrm{d}t}(R^3\dot{x}) = -R^3\frac{\mathrm{d}V}{\mathrm{d}x}.$$

Differentiating, and translating this back into the ϕ notation, the resulting equation of motion is

$$\ddot{\phi} + 3H\dot{\phi} + V'(\phi) = 0, \tag{8.2}$$

where $V'(\phi) = \mathrm{d}V(\phi)/\mathrm{d}\phi$ and $H = \dot{R}/R$ is the usual Hubble parameter. If we knew $H(t)$, this equation could be solved to obtain the change in energy density of the field with time. To find $H(t)$, however, we have to solve Einstein's equations to determine the gravitational effect of the scalar particles. For this it is not quite sufficient to have just the energy density of the field. We know that in relativity the pressure contributes to gravity as well as the energy density. We therefore need to determine the equation of state of the field.

8.4.2 Equation of state

As in chapter 5 we obtain the equation of state from the energy conservation law (5.26) in the form

$$\frac{\mathrm{d}\rho}{\mathrm{d}t} + 3H(\rho + P/c^2) = 0. \tag{8.3}$$

In the present case $\rho = \rho_\phi$ is the mass (or energy) density of the scalar field ϕ. From (8.1) we know the energy density in a comoving volume is

$$\rho_\phi = \tfrac{1}{2}\dot{\phi}^2 + V(\phi), \tag{8.4}$$

so its time derivative is

$$\dot{\rho}_\phi = \ddot{\phi}\dot{\phi} + V'(\phi)\dot{\phi} = -3H\dot{\phi}^2. \tag{8.5}$$

Equation (8.3) can be written

$$\dot{\rho}_\phi = -3H(\rho_\phi + P/c^2) = -3H(\tfrac{1}{2}\dot{\phi}^2 + V(\phi) + P/c^2).$$

Comparing this with (8.5) we obtain the equation of state

$$\frac{P}{c^2} = \frac{1}{2}\dot{\phi}^2 - V(\phi). \tag{8.6}$$

As an example, suppose that $V(\phi) = 0$. Then $P = c^2\rho_\phi$ and we have the case of stiff matter. As a second example, and rather more usefully for inflation, suppose we are given that at some cosmic time, $\dot{\phi} = 0$. Then from (8.4) and (8.6)

$$P = -c^2\rho_\phi. \tag{8.7}$$

Of course, the equation of motion implies that at later times $\dot{\phi} \neq 0$ (because $\ddot{\phi} \neq 0$, unless $\phi(0)$ is at a minimum of the potential), so in this case the equation of state takes this simple explicit form only at one time. In fact, strictly we do not have an equation of state in general, since, except in special cases, we cannot solve for P in terms of ρ_ϕ only. However, we do have an explicit expression for the energy density and pressure of the field which is sufficient for Einstein's equations. Note that the field gives rise to a negative pressure, or a tension, which will remain negative for a time during the evolution (since the evolution is continuous).

8.4.3 Slow roll

Assume that the early Universe is dominated by a scalar field, so that for the moment we can ignore all other forms of matter. We have the equations of motion for the field (8.2) and (8.7) and for the expansion parameter, the Friedmann equation (5.25). In principle, therefore, we have all the ingredients to solve for the scale factor as a function of time. However, to gain some insight into how this model behaves, assume that it starts from conditions in which $\phi = 0$, $\dot\phi \approx 0$. While the second condition holds the equation of state (8.7) is approximately valid, and the energy equation (8.4) implies that $\rho_\phi \approx V_0 = $ constant. This is, approximately, the de Sitter (or steady-state) Universe in which $R(t)$ increases exponentially from some initial value R_i:

$$R(t) = R_i \exp\left(\sqrt{\frac{8\pi G V_0}{3}}\, t\right). \tag{8.8}$$

Therefore, as long as $\dot\phi \approx 0$, the scalar field gives rise to inflation. The main result of this discussion is therefore to justify our use of the de Sitter model as an illustrative example of inflation.

If inflation stops at a time t_f when $R = R_f$, for example because $\dot\phi$ is no longer negligible, the Universe inflates by an amount R_f/R_i. This is usually written

$$\frac{R_f}{R_i} = e^N,$$

where N, the number of e-foldings of the scale factor, gives the amount of inflation. According to (8.8) this is controlled by the value of the potential at $\phi = 0$, and by the time $t_f - t_i$ during which $\dot\phi \approx 0$. How much inflation do we get? From the definition of H we have

$$\frac{\dot R}{R} = H$$

which can be solved for H a general function of t to give

$$\ln\left(\frac{R_f}{R_i}\right) = \int_{t_i}^{t_f} H \, dt.$$

So the number of e-foldings of inflation are

$$N = \int_{t_i}^{t_f} H \, dt = \int H \frac{d\phi}{\dot\phi} = -\int \frac{3H^2}{V'(\phi)} \, d\phi \sim \int \frac{GV}{V'} \, d\phi,$$

where the penultimate equality uses equation (8.2) and the final equality comes from the Friedmann equation with $\rho = V$. The number of e-foldings can be large for V'/V small, hence for a flat potential. In particular models N can easily be of order 100.

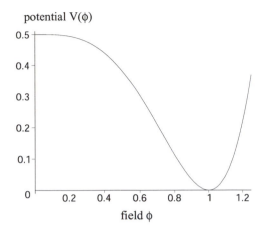

Figure 8.2. A potential of the Coleman–Weinberg form $V(\phi) = 0.5 + \phi^4 (\ln \phi^2 - 0.5)$ plotted against ϕ as an example of a potential with a gentle slope at the origin and a sharp well.

Intuitively therefore we need to start from a large vacuum energy density, and the potential that gives rise to it must have a shallow slope, V', which must remain small as ϕ evolves in order to keep $\ddot{\phi}$, and hence $\dot{\phi}$, small. More precisely, $\dot{\phi} \approx 0$ as long as $\ddot{\phi}$ is negligible in (8.2), hence if $|\ddot{\phi}| \ll |V'(\phi)|$. As long as this condition is satisfied, the equation of motion gives $V'(\phi) \approx -3H\dot{\phi}$, which we can use to eliminate $\dot{\phi}$. Differentiating this gives $V''(\phi)\dot{\phi} \approx -3H\ddot{\phi}$, which we can use to substitute for $\ddot{\phi}$. So, for the potential energy to dominate the kinetic energy, we need

$$1 \gg \frac{\frac{1}{2}\dot{\phi}^2}{V} = \frac{V'^2}{18H^2 V} = \frac{V'^2}{48\pi G \rho_\phi V} \sim \frac{V'^2}{GV^2},$$

where we have put $8\pi G\rho/3H^2 = 1$ for a flat Universe ($\rho = \rho_c$) and we have used $\rho_\phi \sim V$ if $\dot{\phi}$ is small. For the potential energy to continue to dominate or, equivalently for $\dot{\phi}$ to remain small, we need

$$1 \gg \left| \frac{\ddot{\phi}}{V'} \right| \approx \frac{V''}{GV}.$$

For any potential that satisfies these constraints we obtain inflation.

It is customary to picture the evolution of the scalar field by thinking of ϕ as the x coordinate of a particle which rolls slowly down a shallow potential (figure 8.2).

8.5   Fluctuations

Although we treat it as a classical field, the inflaton is really a quantum particle and therefore subject to quantum fluctuations during inflation. The amplitude of these fluctuations turns out to be of order $\delta\phi = H$, the same for all modes. As these fluctuations cross the horizon they become frozen as classical perturbations and they re-enter after inflation as density perturbations. Thus the amplitude at re-entry depends on the amount of growth while outside the horizon. It turns out that this is determined by the constancy of the ratio $\delta\rho/(P + \rho)$. At re-entry $P+\rho \sim \rho$, so the ratio is $\delta\rho/\rho$ evaluated on the Hubble sphere. But, previously, at exit $P_\phi + \rho_\phi = \dot{\phi}^2 \ll \rho_\phi$ so the ratio is $\delta\rho/\dot{\phi}^2$. Thus, equating these expressions, we see that there is significant growth before re-entry, and we find

$$\left(\frac{\delta\rho}{\rho}\right)_H \sim \frac{V'(\phi)\delta\phi}{\dot{\phi}^2} \sim \frac{H^2}{\dot{\phi}}, \tag{8.9}$$

taking $\rho_\phi = V(\phi)$ and $V' = -3H\dot{\phi}$ during inflation.

The right-hand side of (8.9) is to be evaluated when the mode in question exited the horizon, so it is not exactly constant. However, the modes of any interest for the cosmic background radiation and galaxy formation span a range of a factor of order at most a few thousand, or about e^8, in size, hence are of the horizon size for about eight e-folding times out of 100 or so. Therefore, to all intents and purposes the length scales of interest exit the horizon at approximately the same stage. It is therefore a prediction of any inflation model that $(\delta\rho/\rho)_H$ is close to constant on all length scales. More detailed calculations give values for the small, model-dependent departures from constancy known as 'tilt'.

8.6   Starting inflation

In order for inflation to start the inflaton field ϕ must be displaced from equilibrium at $\phi = 0$ towards a new equilibrium at $\phi \neq 0$. The natural way for this to occur is through a change of phase with temperature. Consider the sequence of potentials shown in figure 8.3. As the temperature T decreases the minimum of the potential shifts from $\phi = 0$ to $\phi = \phi_0$. At just above the critical temperature T_c the global minimum shifts to ϕ_0, but there is a barrier between this and local minimum at $\phi = 0$. In this phase transition the thermodynamic quantities are discontinuous and the transition is said to be first order. In particular the discontinuity in free energy is manifested as a latent heat. The ϕ-field can reach the new equilibrium either by thermal fluctuations that get it over the barrier, or by quantum tunnelling. In either case the transition proceeds through the formation of bubbles of the new vacuum. Within a bubble the Universe undergoes inflation. Nevertheless, each bubble is only a small part of the present Universe, which must therefore result from coalescence of bubbles in this version of the theory. We shall see that this presents problems for the end of inflation.

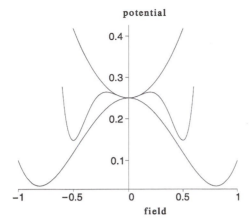

potential

Figure 8.3. Three curves of potential $V(\phi)$ against the field amplitude ϕ showing the transition from a minimum ('false' vacuum) at $\phi = 0$ to the 'true' vacuum at $\phi = 0$.

Alternatively, the potential may pass through a sequence of changes of the form of figure 8.3. The key point is that the $\phi = 0$ minimum disappears at the transition. In this case the transition is continuous and said to be of second order. For a weakly first-order or second-order transition inflation takes place on the same timescale as the phase transition. This can ensure that it involves a patch large enough to encompass the present Universe, thereby dispensing with bubble collisions.

8.7   Stopping inflation

The original idea of Guth (1981) corresponded to a particle rolling down a potential that changes as the temperature falls in such a way that a new minimum appears at $\phi \neq 0$ as the temperature goes below a critical temperature T_c. Each region of the Universe is assumed to be stuck at the original minimum, so supercools until finally, by tunnelling or thermal fluctuations, making a transition to the new minimum. This leaves local bubbles of the symmetry broken phase which, according to the original theory, were supposed to coalesce to a uniform state. The latent heat of the first-order phase transition is eventually released, thereby reheating the Universe.

This cannot work because the slow nucleation rate of bubbles goes hand in hand with the exponential expansion while the Universe is trapped in the false vacuum state. This gives a very inhomogeneous final state. As we stated earlier, the solution turns out to be to invoke a second-order phase transition. This produces domains of true vacuum rather than bubbles surrounded by false vacuum. A single domain can encompass the whole Universe (and more).

In this so-called 'new inflation', the potential is taken to have a steep minimum at $\phi = \sigma \neq 0$ (see figure 8.2). This is the situation of broken symmetry which prevails at normal temperatures. However, following inflation the ϕ field reaches the new minimum of the potential at $\phi = \sigma$ with non-zero $\dot{\phi}$. In fact, all of the initial potential energy will have been turned into kinetic energy. The field therefore oscillates in the potential well around $\phi = \sigma$. If the Universe contained only scalar fields that would be the end of the story. But the scalar field can be assumed to interact with ordinary matter. This coupling damps the oscillations of the ϕ field while exciting those of all the fields to which it couples, i.e. it produces particles of normal matter. (This is familiar in the case of photons which are oscillations of the electromagnetic field. It is also true for all other particles which, quantum mechanically, correspond to oscillations of their associated fields.) If this were not the case then inflation would produce a virtually empty Universe.

Usually one assumes that the ϕ decays to much lighter particles which are therefore relativistic. The ϕ decays produce matter which will, in general, be thermalized. We can therefore discuss the transition in terms of a reheat temperature T_{rh}. This temperature, or the entropy associated with it, is important mainly for compatibility with baryogenesis. If this occurs, as we have described, through $X\bar{X}$ decays, then the reheat temperature must exceed $2M_Xc^2/k$ in order to produce $X\bar{X}$ pairs. An alternative is that baryon asymmetry is produced directly by ϕ decays, or later in the electroweak phase transition, in which case this constraint can be circumvented.

To estimate the reheat temperature there are two cases to consider. First, suppose that the ϕ particle has a large decay width, which means that ϕ particles decay in a time short compared with the expansion rate $(GV(0)/c^2)^{-1/2}$, as ϕ moves to its new minimum of the potential well. In this case all the vacuum energy goes into relativistic particles and the reheat temperature is given by

$$aT_{\mathrm{rh}}^4 = V(0). \tag{8.10}$$

This will also be the reheat temperature for any other mechanism that brings about efficient reheating.

The other case of slow decay occurs if the decay timescale Γ_ϕ^{-1} is comparable to the expansion timescale or longer. Suppose the ϕ field is the dominant matter contribution, any other initial matter having been diluted by inflation. By the time it gets to oscillate in the potential well, the Universe has undergone expansion by a factor e^{100} or so. It can therefore be taken to be cool, so the ϕ particles are themselves non-relativistic. Thus $R(t) \propto t^{2/3}$ as usual. Suppose the ϕ particles start to decay at a time t_i and continue for a time $t = \Gamma_\phi^{-1} \gg t_i$, the decay energy going into relativistic matter. The energy produced will be roughly $V(0) \times (R(t)/R(t_i))^{-3}$, where the final factor allows for dilution as a result of expansion of the initial energy $V(0)$ in the ϕ particles. The dilution of the ϕ particles once inflation has ended is the key difference between this case and the case of rapid decay. Thus we get

$$\rho_r \sim V(0)t_i^2/t^2 \sim \Gamma_\phi^2 c^2/G,$$

where for the start time of the oscillations we have used $t_i \sim H^{-1} \sim$ $(GV(0)/c^2)^{-1/2}$. When the oscillations end $t = \Gamma_\phi^{-1}$ and $\rho_r = aT^4$ gives

$$T_{rh} \sim \left(\frac{c^2}{aG}\right)^{1/4} \Gamma_\phi^{1/2}.$$

In this case the reheating depends on the properties of the ϕ, i.e. on the inflation model.

8.7.1 Particle physics and inflation

At a temperature above about 300 GeV the electromagnetic and weak interactions are unified; the different behaviour we observe below 300 GeV results from a spontaneously broken symmetry involving a scalar field, the Higgs field. The theory has passed many experimental tests, although not yet the explicit confirmation of the existence of the Higgs particle, the mass of which is therefore known only within wide bounds, between 100 GeV and 1 TeV. The nature of the phase transition depends on the mass of the Higgs and its couplings to quarks, but is likely to be second order. It may therefore be associated with some inflation. However, the major period of inflation must occur before baryogenesis and we probably need this to occur well before electroweak symmetry breaking at 300 GeV.

Temperatures around 10^{14} GeV are associated with the unification of the electroweak and strong interactions (the Grand Unified Theories or GUTs). The experimental input into this phase is sparse, although the stability of the proton does rule out some models. But the mechanism here is again assumed to be via a spontaneously broken symmetry involving a Higgs field. The potential is of the Coleman–Weinberg form (figure 8.2) with $\lambda \sim 10^{-3}$ and $\sigma \simeq 2 \times 10^{15}$ GeV. At temperature T the potential (or energy density) acquires a small bump with height $\propto T^4$ with a maximum slightly away from zero. At high enough temperature therefore the scalar field can sit in a stable local minimum near $\phi = 0$. In this situation the symmetry associated with ϕ is unbroken. As the system cools the bump in the potential decreases allowing the field to tunnel out of the minimum near $\phi = 0$ and to begin to roll down the hill. This model is ruled out by the size of the fluctuations it predicts (problem 73).

Alternative models involve multiple scalar fields or modifications to the theory of gravity or supersymmetry (symmetry between fermions and bosons) which can be tuned or given a probabilistic interpretation to yield viable models. An alternative approach is to give up on the particle physics until more is known and seek models that do not depend on phase transitions at all. The simplest example is to start the Universe away from equilibrium at $\phi = 0$ in (say) a ϕ^2 potential, and allow it to slow-roll to $\phi = 0$. This is the chaotic inflation model— chaotic because the initial disequilibrium arises from a random distribution of initial values in (effectively) different Universes. This gives rise to the exotic possibility of regions of the Universe still trapped in the false vacuum state which can bud off as new baby Universes.

8.8   Topological defects

The possibility of trapped bits of false vacuum arises in all theories of spontaneously broken symmetry in an expanding background, the role of the expansion here simply being to force the phase transition by cooling the Universe. Such regions of false vacuum are called defects. For example, a complex scalar Higgs field has a phase as well as a magnitude, and the phase difference round a closed path must be a multiple of 2π for the field to be single-valued. A configuration in which the phase change round a closed path is 2π contains a tube along which the phase is undefined, since otherwise the path could be shrunk continuously to a point along which the phase change is zero, which is impossible since the phase cannot change continuously from 0 to 2π. Along the tube the phase can only be undefined if $\phi = 0$ there, i.e. this configuration contains a tube of false vacuum. Such tubes are called cosmic strings and are either closed or infinite. Cosmic strings could be detected through the gravitational effect on the cosmic background radiation and could play a role in galaxy formation through their gravitational effect on other matter.

The analogous point and two-dimensional defects are not so desirable. Two-dimensional defects are called domain walls and arise on surfaces where the phase of the Higgs field changes sign. Point defects appear in configurations where the Higgs phases align towards a point, which acquires all the properties of a magnetic monopole. Whether these occur depends on the details of a given theory, but if they do they would form a significant component of the mass of the Universe and would be detected. Inflation comes to the rescue here by diluting the density of these objects to an insignificant one per Hubble volume. In a sense, spontaneous symmetry breaking cleans up after itself by sweeping the mess to infinity.

8.9   Problems

Problem 68. *For models with the equation of state*

$$p = w\rho c^2 \qquad -1 < w \le 1$$

show that $R \propto t^{\frac{2}{3(1+w)}}$ (see problem 39) and hence that

$$q = \frac{4\pi G\rho}{3H^2}(1 + 3w).$$

Hence show that accelerating models do not have a particle horizon while decelerating models do. What happens if $q = 0$? Hint: See sections 5.19 and 8.1.

Without making assumptions about the equation of state we can show that positive pressure is a sufficient condition for the existence of a horizon. Show this by considering the integral

$$\int \frac{dt'}{R(t')} = \int \frac{dR}{R\dot{R}}$$

and using the Friedmann equation and energy equation to estimate $\dot{R}$ as a function of R. (Recall that at early times the spatial curvature is small compared to the mass density in the Friedmann equation.)

Problem 69. *Show that in an exponentially expanding Universe ($q = -1$) the Hubble sphere is stationary (see section 5.18). Show that it constitutes an event horizon in the sense that events beyond it will never be observable. Show that in this Universe there is no particle horizon. (In a Universe which has undergone a period of normal expansion with $q > 0$, which then enters a period of exponential expansion, particles which have come within the horizon prior to the onset of inflation are swept beyond the Hubble sphere and contact with them is lost. This contact can then be re-established later once a period of normal expansion is resumed.)*

Problem 70. *If Ω is of order unity today, show that, at the Planck time ($t_P \sim 10^{-43}$ s)*

$$|\Omega - 1| \lesssim 10^{-60}$$

and that the radius of curvature, $R = cH^{-1}|\Omega - 1|^{-1/2}$ exceeded the Hubble radius by 30 orders of magnitude (Kolb and Turner 1990, p 266).

Problem 71. *Consider particles that are non-relativistic today. Show that the mass of such particles within the horizon during the radiation-dominated epoch was*

$$M_h \simeq 0.29 g_*^{-1/2}(\Omega_M h^2) \left(\frac{T}{1\,\text{MeV}}\right)^{-3} M_\odot$$

(Kolb and Turner 1990, p 266). Hint: See equation (7.6) and problem 64.

Problem 72. *The easiest way to estimate the expansion factor e^N that inflation must generate to solve the horizon problem is to look at entropies. (a) Show that the present entropy within the observable Universe is $S \sim 10^{88}k \sim 10^{66}\,J\,K^{-1}$. (b) Entropy is conserved during the inflationary expansion but at its termination the inflaton field decays irreversibly and reheats the Universe approximately to a temperature T_f equal to that before the onset of inflation. Show that*

$$S \lesssim (e^N)^3 (c/H)^3 a T_f^3.$$

(c) If $V(0) = 10^{14}$ GeV^4 estimate the temperature T_f. Hint: See equation (8.10). Hence obtain a minimum value for N.

Problem 73. *Show that the number of e-foldings of inflation in the $V(\phi) = -\lambda\phi^4$ model is of order*

$$N \sim \frac{H^2}{\lambda\phi_i^2}$$

from the time at which the field has the value ϕ_i to the end of inflation ($\phi \ll \phi_i$) and hence show that density perturbations in this model are of order

$$\left(\frac{\delta\rho}{\rho}\right)_H \sim \lambda^{1/2} N^{3/2}.$$

Deduce that $\lambda < 10^{-14}$ is required if the fluctuations are to be compatible with the microwave background. This of course amounts to the fine-tuning inflation is supposed to avoid.

Chapter 9

Structure

9.1 The problem of structure

As a consequence of the existence of small-scale inhomogeneity the Universe is now in a state of gross thermal and dynamical disequilibrium. Yet it appears to have begun in equilibrium and will end in equilibrium. One cannot help but wonder if it could not have chosen an easier, if less interesting, route between its beginning and its end. The problem of structure presents us with the task of accounting for the development of this inhomogeneity.

The obvious approach to the problem is to seek to show that the evolution is from an unstable equilibrium to a stable one. Small initial fluctuations, which must occur naturally in some sense to be specified by the theory, might be amplified to produce the observed structure. The force responsible for this amplification is gravity. How precisely this works is a mystery. The simplest way to outline the subject is to start with an overview of what the problems are.

Early theories, which did not include dark matter, envisaged perturbations in matter growing by gravitational collapse in the same way that stars form from the interstellar medium. A large enough, cool enough mass will collapse under its own excess gravity. In an expanding background this growth is slower than the exponential rate one finds for a fixed background, but given a source of early enough perturbations, of sufficient magnitude, lumps will form. The problem seems to reduce to one of initial conditions. Leaving aside the obvious problem that this only shifts the burden of theory to areas of greater ignorance, there are several intrinsic difficulties. Foremost is the presence of the cosmic background radiation. The interaction between the matter content and this background suppresses the growth of perturbations before decoupling, since charged particles cannot move freely through a field of electromagnetic radiation. We must take this into account in working out the amplitude of the fluctuations at decoupling that corresponds to the magnitude of the density fluctuations we see now. This is not entirely straightforward, partly because the evolution depends on how the mixture of radiation and matter is perturbed initially, and partly because the perturbations now are not small, so cannot be treated as small departures from a homogeneous density field. The possibilities include both a top-down picture in which galaxies form from fragmentation of larger structures if the initial perturbations clump both entropy and energy (by perturbing the matter and radiation together) and a bottom-up picture resulting from an initial clumping of entropy but not energy (by perturbing the matter but not the radiation). The upshot, however, is that (very probably) galaxies do not form in the time available from the level of initial fluctuations permitted by the COBE results.

The white knight that rides to the rescue is dark matter. Whatever this is it comes most likely in one of two main varieties, depending on whether it consists of relativistic or non-relativistic particles at the time, z_{eq}, that matter and radiation have equal densities. The former is referred to as hot dark matter (HDM) and the latter as cold dark matter (CDM). The dark matter is also weakly interacting, hence interacts significantly neither with itself, through collisions, nor with the radiation field through electromagnetic forces. It is therefore not prevented from clumping earlier on (although there is a limit on the amount of clumping on small scales brought about by the streaming of dark matter out of such clumps, unrestrained by collisions). Fluctuations in the dark matter start to grow from the time of matter radiation equality, z_{eq}. This allows a growth factor of greater than 10^5, sufficient to produce the present value of $\delta\rho/\rho \sim 10^2$–$10^3$ in clusters from the value of $\delta\rho/\rho \sim 10^{-2}$–$10^{-3}$ at the time of last scattering, which is the maximum compatible with the COBE results on the smoothness of the cosmic background over the sky. There are, however, some more complications. We have to arrange that the spectrum of fluctuations, that is to say, the relative amplitudes of the fluctuations on various length scales, grow into the distribution of galaxy clustering and produce the fluctuations on various scales in the microwave background. Cutting immediately to the chase, the current position appears to be that, if $\Omega = 1$, $\Omega_\Lambda = 0$ then CDM gives too much power on galactic scales if the fluctuations are fitted to the COBE observations on larger scales. On the other hand, the CDM model is consistent if we allow $\Omega = 1$, $\Omega_\Lambda \sim 0.7$. In HDM models structures on large scales form first, but relatively late on ($z \sim 1$) if they are to match the small-scale structure. Thus these models either have difficulty accounting for the observed structure on galactic scales or for the existence of high redshift galaxies.

9.2   Observations

To set the scene we shall begin with Hubble's attempt to find an edge to the distribution of galaxies (in just the same way that Herschel had tried to find a boundary to the distribution of stars). This involves counting galaxies as a function of limiting magnitude. We shall then go on to consider more modern approaches in which counts of galaxies, and of clusters of galaxies, are used to provide statistical information about their distributions on the sky and to catalogue prominent structures. In section 9.6 we shall show how observations of features in the temperature of the cosmic background radiation on various scales also provide constraints on models of galaxy formation.

9.2.1   The edge of the Universe

In chapter 6 we discussed Hubble's test for the geometry of space by plotting the number of galaxies brighter than a given magnitude against magnitude. For a uniform distribution in a Euclidean Universe we found (equation (6.29))

$$\log N(< m) = 0.6m + \text{constant}.$$

One can look at this alternatively as a test for the uniformity of the galaxy distribution (which was Hubble's point of view). To see this most dramatically consider the case that the local distribution has an edge in the same way that the local distribution of stars has an edge. Then we expect significant departures from a straight line for the faintest objects, provided the observations get to high enough magnitudes (i.e. faint enough). Within the limits of his observations, which were correct only to a factor 2 at the faint end, Hubble found no edge down to a visual magnitude $m_V = 19.8$, corresponding to a distance of about $1000h^{-1}$ Mpc (Peebles 1971).

9.3   Surveys and catalogues

To study the uniformity of the galaxy distribution more carefully requires deeper surveys. Hubble's early work was followed by the Shapley–Ames catalogue (1932) giving the coordinates and magnitudes of 1250 galaxies brighter than 13th magnitude over the whole sky and extended later in the *Reference Catalogue of Bright Galaxies* (de Vaucouleurs and de Vaucouleurs 1964). For larger surveys it is impractical to count individual galaxies. The Shane and Wirtanen catalogue (1967) contains about a million galaxies recorded in $10' \times 10'$ cells. Deeper surveys can be made by recording clusters rather than individual galaxies. The Abell catalogue of clusters (Abell 1958) lists 2712 of the richest clusters to a depth of about 600 Mpc over a large fraction of the sky. Clusters in the catalogue are often referred to by their Abell number; for example, the Coma cluster is A1656.

In the 1980s there was a qualitative leap in the quantity and quality of survey data, including for the first time large-scale surveys that included redshift information, giving genuine three-dimensional pictures. Perhaps the most widely known is the Center for Astrophysics (CfA) redshift survey which by 1985 had recorded 1100 redshifts of galaxies to magnitude in blue light $m_B = 15.5$ in a strip of sky including the Coma cluster. Most of the galaxies have redshifts $z < 0.05$. The results show walls of galaxies concentrated on the edges of large voids. The 1990s brought another order of magnitude improvement. The largest

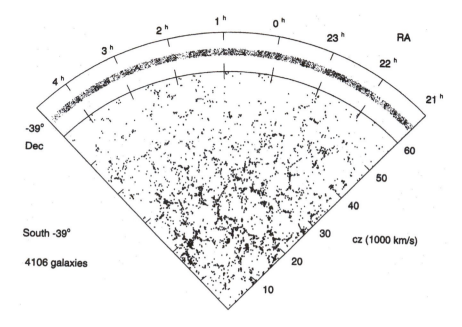

Figure 9.1. Distribution of galaxies in a 3° thick slice of the southern hemisphere from the Las Campanas survey (from Shectman 1996).

survey to date is the Las Campanas Redshift Survey (LCRS, Shectman 1996) which contains about 25 000 galaxies over 700 deg² in six strips with median redshift around 0.1 (figure 9.1). It confirms the relatively nearby voids and walls but appears not to reveal any structure on larger scales (see later). Two major new surveys are currently under way. The Sloan survey (Margon 1999) aims to catalogue over 100 million galaxies in the northern hemisphere with redshifts measured for about a million out to $z \sim 0.2$. In the southern sky the 2DF survey will provide high quality spectra and redshifts for more than 250 000 galaxies. Preliminary results from this survey based on 100 000 galaxies reveal no further large-scale structures.

9.4   Large-scale structures

One can look at the results of the major galaxy surveys in two ways: either as a way of delineating the large-scale structures or as data for a statistical analysis. The major achievements of the searches for structure have been:

(i)   the discovery of the voids, cubic megaparsec volumes that contain less than 10% of the number of galaxies corresponding to a uniform distribution; and

(ii)   the discovery of superclusters, groups of rich clusters of galaxies often in linear or flattened structures.

The early redshift surveys showed linear structures which appear to line up on our galaxy (the 'Fingers of God'). These represent projection effects of our motion relative to the isotropic background (like driving through falling snow). For deeper surveys our motion is, of course, negligible.

This gives the impression that galaxies are distributed in a hierarchical structure of groups, clusters, superclusters and, perhaps, groups of superclusters. These structures are real but not sufficiently prominent to appear as features in the statistical description, to which we now turn.

9.5   Correlations

No theory can hope to reproduce the exact pattern of structures we find in the galaxy distribution. We therefore need to describe the patterns statistically and compare statistics. The simplest guess would be that galaxies are distributed at random. In that case the clusters and voids would be random fluctuations and, on average, galaxies would be no more likely to be found near other galaxies than anywhere else. This turns out not to be the case. Hence we are interested in characterizing the departure of the galaxy distribution from a random one. The most obvious way to do this is to look at the mean square departure from a uniform density of galaxies. This is the galaxy–galaxy correlation function $\xi(r)$, defined explicitly later. A related way of looking at this is to consider $\delta M/M$ the root-mean-square fluctuations in mass δM in volumes containing a mean mass M, for different mass scales M. In principle, one could also look at higher moments of the galaxy distribution.

In practice it is much easier to measure correlations in angle from the projections on the celestial sphere. If the galaxy sample is drawn from a population homogeneous on large scales then the three-dimensional statistics can be deduced from the two-dimensional data. Finally, as well as correlating galaxies we can look at the distribution of clusters.

9.5.1   Correlation functions

We assume that we are analysing the distribution of objects (galaxies or clusters of galaxies) that can be regarded as point particles and that these are distributed homogeneously on a sufficiently large scale, in accordance with the cosmological principle. In particular, we assume that it is meaningful to assign an average number density. For simplicity we shall also assume here that we are dealing with a static Universe and so neglect the effects of expansion and evolution.

If the average number density of galaxies is $\bar{n}$ then we expect, on average, to go a distance $\bar{n}^{-1/3}$ from a given galaxy before we encounter another. Equivalently we expect to find one more galaxy within a distance $\bar{n}^{-1/3}$. We can describe the departure from uniformity by specifying the number that we actually find within this distance. If we were to specify this for each and every galaxy this would, of course, be equivalent to giving the positions of all galaxies, which is not at all the statistical information we want. Rather, we need to take an average. We should therefore imagine many Universes each constructed according to the same statistical law. For each Universe we count up the number of galaxies within a distance r of each galaxy in turn and add up the results. We repeat this for each Universe, and average the result over the ensemble of Universes. In practice, we do not have at our disposal an ensemble of Universes, and we can construct such an ensemble only if we already know the statistical distribution we are trying to determine. The best we can do is to take a spatial average over the visible Universe or as much of it as has been catalogued. This makes sense if we have explored a large enough region of the Universe to encompass many sub-samples representative of the Universe as a whole. For example, if we were to count N galaxies in a random distribution then fluctuations between the ensemble would be of order $N^{1/2}$ which is entirely negligible relative to N for large N. If the Universe were not random, even on the largest visible scale, fluctuations would be of order unity and would not tell us a great deal. We shall assume that the currently visible Universe provides a large enough sample although there is still some debate as to whether this condition is really fulfilled (Wu *et al* 1999).

For a completely random homogeneous distribution of galaxies the probability dP_1 of finding a galaxy in an infinitesimal volume dV_1 is proportional to dV_1 and to the average number density of galaxies, $\bar{n}$, and is independent of position:

$$dP_1 = \frac{\bar{n}}{N} dV_1,$$

where N is the total number of galaxies in the sample. To determine this probability we would divide the galaxy sample into small cells of volumes δV_1 and count the ratio of those cells which contain a galaxy to the total number of cells. The probability of finding two galaxies in a cell is of order δV_1^2 so can be ignored for small enough cells. This counting makes sense if galaxies are distributed uniformly on some scale less than that of the sample. (It would not make sense, for example, if the result were to depend on the sample size.)

If galaxies were not clustered the probability dP_{12} of finding a galaxy in each of volumes dV_1 and dV_2 would, on average, be the product $dP_1 dP_2$ of the independent probabilities for each separately. Any clustering that results not from random fluctuations but from a departure from a random distribution will show up as a departure of the joint probability from a simple product. This defines the two-point correlation function $\xi(r_1, r_2)$:

$$dP_{12} = \frac{\bar{n}^2}{N^2}[1 + \xi(r_1, r_2)] dV_1 dV_2.$$

The assumption of homogeneity implies that ξ depends on the separation $|r_1 - r_2|$ only, not on the location of a pair of galaxies. The assumption that the distribution is random on large scales implies that $\xi(r) \to 0$ as r becomes sufficiently large. Clearly ξ must lie in the range $-1 < \xi < \infty$. A positive ξ implies a tendency of galaxies to cluster together, a negative ξ a tendency towards mutual avoidance.

If ξ is positive for small separations it must become negative as the separation increases beyond some value in order to maintain the average density.

Note that the probabilities and the correlation function are obtained as averages over many galaxies and galaxy pairs. It therefore makes sense to ask about the underlying probability distributions that give rise to these averages. We discuss this in appendix A, p 200.

9.5.2   Linear distribution

Let N galaxies be distributed on a line in non-overlapping clumps of length a, with average density $\bar{n}$ and constant density n_c within the clumps. Let the clumps be distributed at random on the line. This is a model for the situation in which all galaxies occur in clusters which are themselves randomly distributed. We can work out $\xi(r)$ as follows. Pick a galaxy at random. If $r \gg a$ the second galaxy is in another cluster randomly distributed with respect to the first, so $\xi = 0$. If $r \ll a$, the second galaxy will lie in the same cluster, so the probability of finding it in dV_2 is $n_c\, dV_2/N$. Thus

$$\frac{dP_{12}}{dV_1\, dV_2} = \frac{\bar{n}n_c}{N^2} = \frac{\bar{n}^2}{N^2}(1 + (n_c - \bar{n})/\bar{n}),$$

and hence $\xi(r) = (n_c-\bar{n})/\bar{n}$. For r of order a these two limits must join smoothly. (In fact, since the clusters are non-overlapping, for r close to a the galaxies are anticorrelated and ξ must be negative.) The main point is that the clustering on a scale a introduces a 'knee' in the correlation function. Conversely, the presence of such a knee in the observed data would indicate a characteristic scale of clustering. This is not what we shall find.

9.5.3   The angular correlation function

To obtain the spatial correlation function $\xi(r)$ directly requires three-dimensional positional information on the galaxy distribution. This information requires extensive observation of redshifts and hence large amounts of telescope time. Until relatively recently, therefore, all large galaxy catalogues recorded positional information on the sky without reference to depth. Thus we can obtain directly an angular correlation function, defined in appendix B, p 202. From this $\xi(r)$ can be reconstructed if we assume isotropy and homogeneity. It has now become possible to measure large numbers of redshifts simultaneously, for example around 600 at a time in the Sloan survey, from which $\xi(r)$ can be obtained directly.

9.5.4   Results

Figure 9.2 shows the galaxy–galaxy correlation function $\xi(s)$ plotted against separation s for a combination of redshift surveys and compared with the function

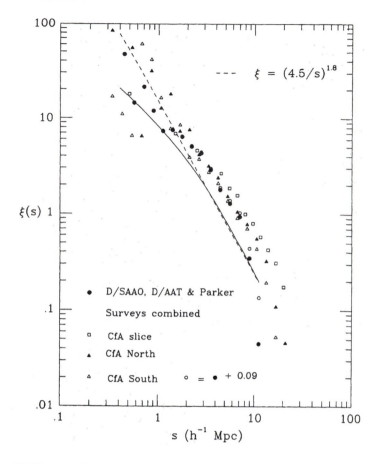

Figure 9.2. The correlation function $\xi(s)$ from several redshift surveys. (The separation s deduced from the recession velocity is used rather than the distance r as the abscissa to emphasize that the data are not corrected for local velocities.) From Shanks *et al* (1989).

$$\xi_{gg}(s) = \left(\frac{s}{r_g}\right)^{-1.8}, \qquad (9.1)$$

where r_g is of order $4.5h^{-1}$ Mpc. Strictly the separation, estimated from the redshifts in these surveys, is distorted from the Hubble flow by the motions induced by the galaxy clustering we are measuring. For the present purposes this can be ignored, so $s \approx r$. The results are consistent with measured angular correlation functions to various depths. The power law form (9.1) holds from $0.1h^{-1}$ Mpc to $10h^{-1}$ Mpc beyond which the correlation function drops rapidly to zero beneath the noise in the data.

The correlation of clusters in the Abell catalogue gives the cluster–cluster correlation function

$$\xi_{cc}(r) = \left(\frac{r}{r_c}\right)^{-1.8}$$

between about $5h^{-1}$ Mpc and $75h^{-1}$ Mpc with r_c of order 12–$25h^{-1}$ Mpc. We have

$$\xi_{cc} = \left(\frac{r_g}{r_c}\right)^{-1.8} \left(\frac{r}{r_g}\right)^{-1.8} \simeq 6 \times \xi_{gg}.$$

Thus, rich clusters are clustered more strongly than galaxies (i.e. $\xi_{cc} > \xi_{gg}$) (Kaiser 1984).

9.6   Bias

The observation that rich clusters are clustered more strongly than galaxies means that both cannot represent the underlying matter distribution, and perhaps neither does. For example, galaxies might form only above a certain density contrast, hence preferentially at the peaks of the density distribution. It can be shown that the peaks would then be more strongly correlated than the underlying mass fluctuations. This suggests we introduce a biasing parameter b, for the density contrasts $\delta = \delta p / p$

$$\delta_{galaxies} = b\delta_{matter},$$

and that rich clusters are biased towards the peaks of the mass fluctuations (Bahcall 1988, section 3.5).

9.7   Growth of perturbations

The study of galaxy formation is a complex subject. We begin by outlining the various aspects. First we consider the conditions under which a small density enhancement in an otherwise uniform matter distribution can grow under its own gravity. To set the scene this is done first for a static Universe, after which we look at the essential difference introduced by an expanding background in Newtonian gravity. The Newtonian approximation is sufficient to treat a perturbation of the density on a scale smaller than the horizon for which the effect of spacetime curvature can be neglected (see problem 78). Larger scale perturbations are important and we shall quote the results from a relativistic treatment. Gravity is not the only force at work: we consider next the influence of the background radiation which prevents the growth of smaller scale perturbations. In addition, the presence of radiation means we have to define rather more carefully what it is that is perturbed (matter or matter and radiation). This leads us to distinguish two basic types of perturbation (called adiabatic and isocurvature or isothermal) which lead to very different pictures.

The problem is further complicated by the existence of two types of matter—baryonic and dark matter—and by the various possibilities for the dark component. To simplify the picture we shall first explain the theory in the absence of dark matter, as it was developed historically. In this case adiabatic perturbations lead to the formation first of the largest scale structures from which the galaxies form by fragmentation. With isothermal perturbations galaxies arise from mergers of smaller scale fragments.

The presence of dark matter leads to different behaviours because it interacts differently with the microwave background. In the presence of a dominant dark matter component we can consider the baryons to follow the gravitational pull of the dark matter perturbations once they decouple from the background radiation. We consider only adiabatic perturbations, but there are still the two main alternatives in which galaxies form by fragmentation or aggregation. This depends on whether the dark matter is relativistic (hot) or non-relativistic (cold) when it decouples from the radiation.

Galaxies now represent an overdensity relative to the mean that exceeds a factor of two. This means that the late stages of galaxy formation cannot be treated as small perturbations using the linearized theory. Instead, one has to find a way of incorporating the full nonlinear behaviour of the collapse of overdense regions; but this nonlinear growth is beyond our scope.

At this stage we have considered the evolution of perturbations only on a single, although arbitrary scale. We shall find that perturbations on different scales have different growth rates. To complete the picture therefore, we discuss the usual assumptions about the spectrum of perturbations in the early Universe, i.e. the amplitude of the initial fluctuations on each length scale. The intention is, of course, that physical processes acting on the input spectrum should lead to the observed statistical properties of the galaxy distribution.

9.7.1 Static background, zero pressure

Suppose a small irregularity occurs in a uniform static self-gravitating medium. If pressure forces can be neglected an overdense region will tend to increase in density further as a result of its increased self-gravity, and will continue to grow until pressure forces do intervene. We show how, in Newtonian physics, this positive feedback leads to an instability characterized by exponential growth.

The gravitational potential ϕ at each point is given by Poisson's equation

$$\nabla^2 \phi = 4\pi G \rho, \tag{9.2}$$

while the velocity field u induced in the medium is controlled by the Euler equation of motion

$$\frac{\partial u}{\partial t} + u \cdot \nabla u = -\nabla \phi, \tag{9.3}$$

and the conservation of mass equation for the density ρ

$$\frac{\partial \rho}{\partial t} + \nabla \cdot (\rho \boldsymbol{u}) = 0. \tag{9.4}$$

Now imagine a large static region of uniform density ρ_0 and constant gravitational potential ϕ_0. (Strictly, you should find yourself unable to do this, because such a distribution of matter under its own gravity cannot remain static, but would collapse; ignore this because it is not important for the development of this introductory argument.) Consider a small perturbation of the density to $\rho = \rho_0 + \rho'$, of the velocity from zero to $\boldsymbol{u}'$ and of the gravitational potential to $\phi_0 + \phi'$. Linearizing the equations (9.2)–(9.4) in the small quantities ρ', $\boldsymbol{u}'$ and ϕ' gives

$$\nabla^2 \phi' = 4\pi G \rho',$$

$$\frac{\partial \boldsymbol{u}'}{\partial t} = -\nabla \phi',$$

$$\frac{\partial \rho'}{\partial t} + \rho_0 \nabla \cdot \boldsymbol{u}' = 0.$$

Eliminating $\boldsymbol{u}'$ and ϕ' gives

$$\frac{\partial^2 \rho'}{\partial t^2} = (4\pi G \rho_0)\rho'. \tag{9.5}$$

Equation (9.5) tells us that the perturbation in density grows exponentially on a timescale $(4\pi G \rho_0)^{-1/2}$. The system is unstable in that any small fluctuation will grow to a significant size in a finite time. When a corresponding analysis is performed to determine the fate of a density fluctuation in a uniform expanding Universe the result is significantly different, as we now show.

9.7.2 Expanding background

To illustrate the behaviour of perturbations in relativistic cosmology we consider a special case. Inside a given spherical region we imagine the matter to have been compressed slightly to a higher, but still uniform density. So our perturbed Universe contains a spherical mass M of density ρ_1 surrounded by an empty shell surrounded by the unperturbed Universe of density ρ_0. (Note that ρ_0 refers to the unperturbed background in this section, not the present time.) Since the mass M is constant in time the scale factor R_1 of the inner sphere satisfies

$$\tfrac{4}{3}\pi R_1(t)^3 \rho_1 = M = \text{constant}, \tag{9.6}$$

and, since the evacuated region is part of an FLRW model with scale factor R_0,

$$\tfrac{4}{3}\pi R_0(t)^3 \rho_0 = M. \tag{9.7}$$

The Friedmann equations (5.44) for the two scale factors are

$$\frac{1}{2}\dot{R}_i^2 - \frac{GM}{R_i} = E_i = \text{constant} \tag{9.8}$$

where $i = 0, 1$, and we have used (9.6) and (9.7) to eliminate the densities.

Let δ be the fractional fluctuation in density,

$$\delta = \frac{\rho_1 - \rho_0}{\rho_0}.$$

Using (9.7) $-\delta/3$ also gives the fractional difference in scale factors:

$$-\frac{1}{3}\delta = \frac{R_1 - R_0}{R_0}.$$

Assuming $\delta \ll 1$, we expand (9.8) to first order in δ and differentiate with respect to t to get

$$\ddot{\delta} + 2H\dot{\delta} - 4\pi G\rho_0\delta = 0, \tag{9.9}$$

where $H = \dot{R}_0/R_0$ in the unperturbed background. This determines the evolution of a density perturbation.

To see the effect of expansion consider as an example the Einstein–de Sitter model with $6\pi G\rho_0 = t^{-2}$. Solving (9.9) in this case, using the trial solution $\delta \propto t^p$ gives

$$\delta = At^{2/3} + Bt^{-1},$$

where A and B are arbitrary constants. The decreasing mode Bt^{-1} is not important since it dies out in time. The key point is that density fluctuations do grow, but that the growth of the increasing mode is no longer exponential, but a power law: against an expanding background perturbations grow more slowly. This conclusion is valid for other cosmological models and more general initial perturbations. Thus, compared with the static model, in order to grow to a given present density contrast a fluctuation must have been much larger at a given time in the past, and, equivalently, a fluctuation of given amplitude must have occurred much earlier.

9.8 The Jeans' mass

Gravity provides an amplification mechanism but amplifiers by themselves do not provide structure. If we are to do better than simply attribute all structure to initial conditions then some forces other than gravity must be operating, since gravity alone provides no characteristic length scale. There are two questions to be asked: can the initial spectrum of fluctuations be predicted and how is that spectrum modified over time?

Consider then the effect of pressure forces which we previously omitted, but neglect expansion. The Newtonian equations become

$$\nabla^2\phi = 4\pi G\rho,$$

$$\frac{\partial u}{\partial t} + u \cdot \nabla u = -\nabla\phi - \frac{1}{\rho}\nabla P,$$

$$\frac{\partial \rho}{\partial t} + \nabla \cdot (\rho u) = 0.$$

To complete the prescription we have an equation of state $P = P(\rho)$ which we can leave unspecified at the moment. We perturb these equations away from the static solution as before, but now with a small perturbation to the pressure $P' = (\mathrm{d}P/\mathrm{d}\rho)\rho' = c_s^2\rho'$, where c_s will turn out to be the speed of sound in the medium. This yields

$$\frac{\partial^2 \rho'}{\partial t^2} = (4\pi G\rho_0)\rho' + c_s^2\nabla^2\rho',$$

a wave equation for ρ' with wave speed c_s (compare with equation (9.5)). We therefore look for plane wave solutions of the form

$$\rho' \propto \mathrm{e}^{\mathrm{i}(\omega t - k \cdot x)}.$$

Substitution yields the dispersion relation

$$\omega^2 = -4\pi G\rho_0 + c_s^2|k|^2.$$

For $c_s^2|k|^2 < 4\pi G\rho_0$ we have, as in the pressure-free case, $\omega^2 < 0$. Thus ω is imaginary and one of the solutions for ρ' grows exponentially. On the other hand, for $c_s^2|k|^2 > 4\pi G\rho_0$ the solutions for ω are real and the density perturbation oscillates. Since $P' \propto \rho'$, the pressure tracks the density: and the disturbance in this case propagates as a sound wave through the medium with speed c_s. Pressure therefore acts as a stabilizing force on scales

$$\lambda_J = \frac{2\pi}{|k|} < \left(\frac{\pi}{G\rho_0}\right)^{1/2} c_s.$$

On larger scales gravity dominates and the perturbations are unstable. The critical scale λ_J is called the Jeans' length. The mass in a sphere with radius $\lambda_J/2$ is the Jeans' mass

$$M_J = \frac{\pi}{6}\left(\frac{\pi}{G\rho_0}\right)^{3/2} c_s^3.$$

Although it is beyond the scope of our treatment to demonstrate the fact, roughly the same Jeans' length is found in expanding Newtonian models and in relativistic models, although in the latter case this is not the whole story since it turns out that there are additional modes that can be excited corresponding to gravitational waves and vortical motions of the medium.

9.9   Adiabatic perturbations

Since the real Universe contains matter and radiation it is not a single fluid, so the previous analysis requires some adaptation. In particular we need to consider in more detail the nature of the perturbations. We shall treat the dark matter component later. For the present we consider only baryonic matter and radiation. The significance of the distinction is that baryonic matter interacts electromagnetically, whereas the dark matter is only weakly interacting hence does not 'see' the radiation field. To simplify still further we ignore the distinction between the decoupling of matter and radiation and the transition from radiation to matter dominance and take these to have occurred at a common redshift z_{eq}.

A general perturbation can be decomposed into the sum of two parts which behave differently. These are curvature perturbations, dealt with here, and isocurvature perturbations treated in the next section. For our current considerations, we take the energy density to be made up of matter, $\rho_m c^2$ and electromagnetic radiation $\rho_r c^2$. In a curvature perturbation the total energy density $(\rho_r + \rho_m)c^2$ is perturbed but not the entropy per baryon s. Changes in energy density give rise to changes in curvature by Einstein's equations. The entropy in a comoving volume $S = \frac{4}{3}aT^3V$ is conserved (since $T \propto 1/R$ and $V \propto R^3$); hence curvature perturbations are adiabatic.

Furthermore, $\delta \log s = 0$ implies

$$3\frac{\delta T}{T} - \frac{\delta \rho_m}{\rho_m} = 0, \tag{9.10}$$

and, since $\rho_r = aT^4$,

$$\frac{\delta \rho_r}{\rho_r} = \frac{4}{3}\frac{\delta \rho_m}{\rho_m}.$$

Equation (9.10) tells us how adiabatic perturbations in the matter are reflected in perturbations in the temperature of the cosmic background radiation in the case that the matter is purely baryonic. (The perturbation in the radiation field is significantly less by one or two orders of magnitude in the case that the dominant matter component is dark.)

We are interested in the Jeans' mass for an adiabatic perturbation at recombination. To avoid the complication of the transition from radiation-dominated to matter-dominated evolution, we calculate the Jeans' mass at matter–radiation equality, z_{eq} and ignore the period (if it exists) between this and recombination. We have $\rho \approx \rho_r$ since $\rho_m \sim \rho_r \sim \rho_{eq}$ and $P \approx P_r \approx \frac{1}{3}\rho_r c^2$. In this situation the speed of sound $(dP/d\rho)$ is $c/\sqrt{3}$ and the Jeans' mass is

$$M_J^{(a)}(z_{eq}) \simeq \frac{\pi}{6}\rho_{eq}\left(\frac{c}{\sqrt{3}}\right)^3 \left(\frac{\pi}{G\rho_{eq}}\right)^{3/2}$$

$$\simeq 3.5 \times 10^{15}(\Omega_M h^2)^{-2} M_\odot.$$

(see section 5.16 for ρ_{eq}). This is the mass of a supercluster of galaxies. We shall see that this is not quite the whole story, but it suggests that pure adiabatic perturbations are associated with the formation initially of structures on super-galactic scales: a 'top-down' approach to galaxy formation. Note, however, that this conclusion does not carry over to the case when dark matter is present.

9.10 Isocurvature (isothermal) perturbations

In the second type of perturbation the entropy is changed but not the energy density. This means that the curvature remains unchanged, because Einstein's equations relate it to the energy density. But the equation of state is varied locally, because that ratio of matter energy to radiation energy changes. From $\delta\rho = 0$ we have $\delta\rho_r = -\delta\rho_m$ and hence for the temperature

$$4\frac{\delta T}{T} = \frac{\delta\rho_r}{\rho_r} = -\frac{\delta\rho_m}{\rho_r} = -\frac{\rho_m}{\rho_r}\frac{\delta\rho_m}{\rho_m}.$$

But in the radiation era $\rho_m/\rho_r \ll 1$, so $\delta T/T \approx 0$. Thus, isocurvature perturbations are also approximately isothermal (and the two terms are often used interchangeably). In an isothermal perturbation the matter density is perturbed locally, but not the radiation density, or, what amounts to the same thing, there is a perturbation in the local equation of state of the combined fluid.

In this situation the sound speed is $dP_m/d\rho_m$ (because the radiation pressure P_r is being kept constant) which is just the usual expression for the sound speed in a gas. Hence, at recombination when the gas is atomic hydrogen, $c_s = \sqrt{(\gamma k T_{eq}/m_p)} \simeq 5 \times 10^5$ m s^{-1} for $T_{eq} \simeq 4000$ K, again neglecting the period between matter–radiation equality and recombination. The Jeans' mass at recombination for an isothermal perturbation is

$$M_J^{(i)}(z_{eq}) \simeq \frac{\pi}{6}\rho_{eq}c_s^3\left(\frac{\pi}{G\rho_{eq}}\right)^3$$
$$\simeq 5 \times 10^4 (\Omega h^2)^{-1/2} M_\odot.$$

This suggests that isothermal perturbations provide structures of globular cluster size that grow by aggregation.

After recombination the radiation is irrelevant and the two types of perturbation evolve in the same way. However, we have not yet taken into account two effects. First, we need to consider the role of expansion. The main qualitative influence of this is through the existence of a horizon. Second, during recombination the two fluids, matter and radiation, do not behave as a perfect fluid and we have to take account of dissipation.

9.11   Superhorizon size perturbations

The horizon is at $2ct$ to $3ct$ depending on whether the expansion is radiation or matter dominated. For order of magnitude estimates we take $R_h = ct$. The baryon mass within the horizon just before decoupling is therefore

$$M_h(t_{eq}) = \frac{\pi}{6} \rho_m (ct_{eq})^3$$
$$\simeq 6 \times 10^{13} \Omega_M^{-2} h^{-1} M_\odot.$$

This is less than the Jeans' mass for adiabatic perturbations. This means that a perturbation on the Jeans' scale lies outside the horizon and so cannot be treated by a Newtonian analysis. In fact, a full relativistic treatment shows that the density contrast $\delta\rho/\rho$ for superhorizon sized adiabatic perturbations grows as $\delta\rho/\rho \propto R(t)$ in the matter-dominated regime and as $\delta\rho/\rho \propto R(t)^2$ in the radiation-dominated phase. Superhorizon isocurvature fluctuations do not grow while they remain outside the horizon. This can be understood if we recall that these perturbations correspond to changes in the equation of state, or pressure, keeping the energy density constant. Thus there are no gravitational gradients. Such changes cannot propagate over non-causally connected regions.

9.12   Dissipation

In an adiabatic perturbation photons can diffuse out of an overdense region. Photon viscosity—the fact that the electrons cannot move freely through a radiation field because of Thomson scattering—means that the diffusing photons drag the electrons with them. This smoothes the perturbation on scales less than λ_S, the Silk scale. The corresponding baryonic mass turns out to be of order $M_S = 6 \times 10^{12} (\Omega_M h^2)^{-5/4} M_\odot$, tantalizingly of order of the mass of a cluster of galaxies.

9.13   The spectrum of fluctuations

So far we have considered separately the fate of perturbations on various scales. In fact we expect galaxy formation to start from the initial presence of fluctuations on a range of scales, depending on how the fluctuations are produced. It is then the evolution of this spectrum that is to be compared with the galaxy correlation functions. There are two current mainstream theories for this initial input. These are cosmic strings and inflation. Cosmic strings can be thought of as a particular type of non-thermal relic from the early Universe. Their detailed treatment is beyond our scope and, in any case, they are not favoured at the moment (de Bernardis *et al* 2000). We shall link what we have to say here more closely to the inflation picture of the origin of fluctuations which we discussed in chapter 8.

If we believe that the galaxy correlations arise from the physics of hot gravitating matter in an expanding Universe, rather than from the input conditions, then there should be no length scale in the input spectrum. To see how this works consider again a power law correlation function

$$\xi(r) \propto r^{-\alpha}. \tag{9.11}$$

Suppose we rescale the radial coordinate by a factor a. Then $\xi(r/a) \propto a^{\alpha} r^{-\alpha} \propto \xi(r)$, so the rescaled correlation function is unchanged in shape. Compare this with the example in section 9.5.2 where the knee of the distribution fixed the characteristic length scale. To avoid any possible confusion, note that the scales associated with the power law fit to the observed galaxy clustering are obtained from the amplitude of the correlation function (the length r_0 for which $\xi(r_0) = 1$), not its shape.

It is perhaps more natural to think of fluctuations in mass on each scale rather than the correlation function. We can relate the two as follows. Define

$$J_3(R) = \int_0^R \xi(r) r^2 \, dr;$$

then the mass fluctuations on a scale R can be shown to satisfy

$$\left\langle \left(\frac{\delta M}{M} \right)^2 \right\rangle \sim \frac{J_3}{R^3}.$$

If $\xi(r)$ has the power law form (9.1) then

$$\left(\frac{\delta \rho}{\rho} \right)_R = \left(\frac{\delta M}{M} \right)_R \propto M^{-\alpha/6},$$

where root-mean-square values are understood.

The cosmological principle implies that the mass should not be concentrated in structures on ever larger scales, which turns out to translate to $\alpha > 4$. On the other hand, large density enhancements on the smallest scales would lead to the formation of a population of small black holes that would be evaporating and contributing to the gamma-ray background now. In order not to exceed the gamma-ray limits it turns out that we must require $\alpha \approx 4$. Thus it is usual to assume the so-called Harrison–Zeldovich spectrum

$$\frac{\delta M}{M} \propto M^{-2/3}. \tag{9.12}$$

It can be shown that for perturbations of this form the fluctuation in density $\delta \rho / \rho$ as the perturbation comes within the horizon is the same for all scales:

$$\left(\frac{\delta \rho}{\rho} \right)_H = \left(\frac{\delta M}{M} \right)_H = \text{constant} \tag{9.13}$$

as a mass scale M enters the horizon (see problem 76). The result arises because larger mass scales enter the horizon later, having therefore had more time to grow. In the (unrealistic) case that the perturbation enters the horizon in the matter-dominated era we can obtain this result from our previous discussion. A perturbation on a length scale $1/k$ enters the horizon when its proper length equal the horizon size, hence at a time t_* given by

$$ct_* = R(t_*)/k.$$

In the matter-dominated case we have $R(t) \propto t^{2/3}$, hence

$$t_* \propto k^{-3}.$$

During this time the perturbation grows as $\delta \propto t_*^{2/3}$ (section 7.2) or as $\delta \propto k^{-2} \propto M^{2/3}$, which relates (9.13) and (9.12).

Equation (9.13) gives another way of seeing that there is no intrinsic length scale. Inflation models predict a scale-free spectrum of adiabatic fluctuations of this form.

9.14   Structure formation in baryonic models

We can now summarize the formation of structure in a baryonic Universe.

Adiabatic fluctuations in mass on scales less than the Silk scale M_S are damped by interaction with the radiation field. Only larger scale perturbations can grow. Starting from a Harrison–Zeldovich spectrum of adiabatic fluctuations, by the time we reach decoupling at $\sim z_{eq}$, the fluctuations will be dominated by those on the smallest surviving scale, namely M_S.

For adiabatic fluctuations we have $\delta\rho/\rho \sim \delta T/T$ and we can estimate the $\delta\rho/\rho$ required to produce galaxies today. Thus we can predict the fluctuations in the microwave background. These turn out to be much larger than those observed.

For isothermal fluctuations there is no growth of perturbations outside the horizon and no dissipation. For a Harrison–Zeldovich spectrum with less power on larger scales, the only scale that matters is the Jeans' mass M_J. Galaxies can be built up by mergers on sub-galactic scales. There is no conflict in this case with the smoothness of the microwave background, but it is difficult to know where the fluctuations might come from (see also Kolb and Turner 1990)

Arguments over the merits of the adiabatic or isothermal approaches were, however, overtaken by the appearance of dark matter models. In fact, it is sometimes claimed that the ability of dark matter models to circumvent the problems of galaxy formation in baryonic models is evidence for the existence of non-baryonic dark matter (see section 3.10).

9.15   Dark matter models

The effect of dark matter on structure formation depends upon its temperature at decoupling. Particles that were relativistic when they decoupled from the remaining matter and radiation form the hot dark matter (HDM). Particles that were non-relativistic constitute cold dark matter (CDM). For particles that interact through the weak force, the distinction amounts to one between low mass and massive particles as constituents of the dark matter. For example, neutrinos with a small mass have been proposed as candidates for HDM, while other exotic particles such as axions or neutralinos could be the CDM. If we consider even more exotic particles, allowing the interaction strength to vary, then other cases are possible. For example, a low mass particle (a few keV say) decoupling at a few hundred GeV (instead of a few MeV) would be relativistic without being (relatively) hot. This intermediate case of so-called warm dark matter has not been much explored.

Perturbations in non-baryonic dark matter are not coupled to the radiation field. Thus there is no direct analogue of the dissipation of structure on small scales by photon viscosity. Nevertheless, fluctuations on a small scale are smoothed because the dark matter can stream freely out of them. The Silk scale is, therefore, replaced by the free-streaming scale, which can be estimated as

$$\lambda_{fs} = R(t) \int_0^{t_{eq}} \frac{v(t')}{R(t')} \, \mathrm{d}t',$$

where the integral is taken up to t_{eq} at which perturbations can start to grow. This time encompasses both a relativistic regime, $v \sim c$, and a non-relativistic period with $v \propto R^{-1}$. Evaluating the two contributions for a particle at some temperature T_d gives

$$\lambda_{fs} \sim 30(\Omega h^2)^{-1}(T_d/T)^4 \text{ Mpc},$$

which is about a factor two smaller than the accurate value obtained from the evolution of the particle distribution function (problem 77; see Kolb and Turner (1990)). We shall take $\lambda_{fs} = 40(\Omega h^2)^{-1}$ Mpc translated to the present time.

9.15.1   Growth of fluctuations in dark matter models

Consider first a spectrum of adiabatic perturbations in HDM. Dark matter particles stream out of over-dense regions so damping perturbations on scales less than about $40(\Omega h^2)^{-1}$ Mpc (as measured at the present time). Thus, the first structures to form are of this size, corresponding to superclusters of $10^{15} M_\odot$ or more. Structures on this scale are unlikely to collapse spherically. The likely outcome is therefore a 'pancake' structure. Baryons in this structure collide and dissipate gravitational energy, condensing finally into galaxies. Numerical simulations in this picture reproduce the voids and sheets of galaxies, as might be expected. However, the characteristic length scale of 40 Mpc does not match the nonlinear break in the correlation function on a scale of 5 Mpc. As a consequence, it turns out to be difficult to reproduce the galaxy–galaxy correlation function with reasonable assumptions about the epoch of galaxy formation.

For CDM the damping scale is smaller than 1 Mpc, so the formation of structure follows the initial perturbation spectrum with $\delta\rho/\rho$ increasing to small scales. Thus, the first objects that form on the Jeans' length are sub-galactic. If we assume a flat Universe with zero cosmological constant then numerical simulations give too much structure on cluster scales. Numerical simulations which fit the correlation function require an open Universe with $\Omega h^2 = 0.2$, but fail to produce the large-scale structures and voids. The introduction of bias, or of a non-scale invariant initial spectrum (called tilt in the literature), can redress some of these deficiencies. But no models with a zero cosmological constant can simultaneously satisfy the constraints provided by detailed observations of the microwave background, which promise to yield well-constrained values for the cosmological parameters.

9.16 Observations of the microwave background

Consider a simple adiabatic density perturbation in a baryonic Universe. As we saw earlier, the adiabatic condition requires a constant entropy per baryon, so we have $\delta(\frac{4}{3}aT^3/\rho_m) = 0$, or

$$\frac{\delta T}{T} = \frac{1}{3}\frac{\delta\rho_m}{\rho_m}.$$

Thus the measurement of fluctuations in the background temperature provides information about the matter fluctuations. A similar relation holds for dark matter models with a model-dependent factor (instead of one-third) which is between one and two orders of magnitude smaller.

It is useful to bear in mind a simple relation between the angular scale of a fluctuation on the last scattering surface $\delta\omega$ and the corresponding length scale (measured at the present time):

$$1h^{-1}\ \text{Mpc} \simeq \tfrac{1}{2}\omega\ \text{arcminutes}.$$

It follows that the COBE observations which map the sky at $\sim 10°$ resolution provide information on fluctuations on scales of 360 Mpc or more. These therefore give only indirect evidence of fluctuations on scales relevant to galaxy formation, by extrapolation of a theoretical spectrum.

To characterize the temperature fluctuations we use the angular power spectrum. Expand the departure of the temperature from the mean in any direction labelled by spherical polar angles θ and ϕ in terms of spherical harmonics $Y_{lm}(\theta, \phi)$,

$$\delta(\theta, \phi) \equiv \frac{\Delta T}{T} = \sum_{l=0}^{\infty}\sum_{m=-l}^{m=l} a_{lm}Y_{lm}(\theta, \phi),$$

(e.g. Arfken and Weber 1995).

Then, by analogy with the galaxy distribution, we define the autocorrelation function, which, using properties of spherical harmonics, can be shown to be

$$\langle \delta(\theta, \phi) \delta(\theta', \phi') \rangle = \frac{1}{4\pi} \sum_{l} (2l + 1) \langle |a_{lm}|^2 \rangle P_l(\cos \alpha).$$

The angular brackets as usual denote, in theory, an ensemble average and, in practice, an all-sky average, and α is the angle between the directions (θ, ϕ) and (θ', ϕ'). The function $P_l(\cos \theta)$ is a Legendre polynomial. The average $C_l \equiv \langle |a_{lm}|^2 \rangle$ is called the power spectrum on the angular scale $\sim 2\pi/(l + 1)$, and is independent of m. Putting $\theta = \theta'$, we have

$$\left\langle \left(\frac{\Delta T}{T} \right)^2 \right\rangle = \frac{1}{4\pi} \sum_{l} (2l + 1) C_l.$$

The $l = 0$ term represents a correction to the assumed mean temperature and the $l = 1$ term is the dipole contribution, usually assumed to arise from our motion. The $l = 2$ term is an intrinsic quadrupole. The COBE result for the $l = 1$ term is given in chapter 4.

The higher l terms, which will be measured by the Planck and MAP satellites, provide information on smaller angular scales. In particular, as oscillating perturbations (or sound waves) come within the horizon, some electrons will be moving towards the observer and some away, setting up a fluctuation in temperature at the observer from the relative Doppler shift. The effect for a perturbation of wavelength λ at time t is of order

$$\frac{\delta T}{T} \sim \frac{\delta \rho_m}{\rho_m} \frac{\lambda}{ct},$$

although a full numerical treatment of the radiative transfer through recombination is required to obtain the details. This temperature fluctuation appears as a peak in the power C_l as a function of l. The position of this 'Doppler' peak as a function of l can be estimated as the angle subtended at the observer by the acoustic horizon at decoupling, since this is the scale on which material can be moving coherently. Since the sound speed is of order $c/\sqrt{3}$ at this time, the acoustic horizon is of the same order as the particle horizon. This gives

$$l \simeq 220 \Omega^{1/2},$$

so for $\Omega = 1$ the angle is just less than $1°$. The location of the first peak depends principally on the total Ω and is relatively insensitive to the other parameters, Ω_B, h and Ω_Λ. The most recent measurements of this peak at the time of writing come from the balloon-borne experiments MAXIMA and BOOMERanG. They each give a value for Ω close to unity. The height of the peak measures the relative

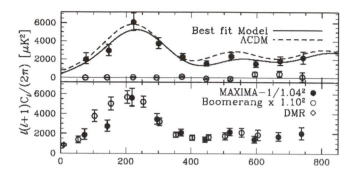

Figure 9.3. Bottom: The power in the microwave background signal in each (spherical) harmonic component as measured by the two balloon-borne experiments MAXIMA-1 (filled circles), and BOOMERanG (open circles). Top: Comparison of the experimental points with predictions from two models having $\Omega_0 = 1$ and containing CDM and a non-zero cosmological constant.

contribution of baryonic matter Ω_B. (In fact the height depends on other things as well, so in practice the heights of two peaks are required.) For Ω_B the data from MAXIMA and BOOMERanG yield 0.032. This is within two standard deviations of the value obtained from nucleosynthesis calculations. Measurements of the peaks of greater accuracy are anticipated from the MAP and Planck satellite experiments.

Detailed numerical models produce additional smaller peaks at higher l (see figure 9.3). The positions of these peaks are sensitive to Ω and to the matter perturbations, but will be difficult to resolve. In addition, they are smeared, for example, by the finite thickness of the last scattering surface.

When the current data on the cosmic radiation background from COBE and balloon observations are put together with determinations of fluctuations on smaller scales from galaxy correlations the best fit model is found to be one which is flat, $\Omega = 1$, and in which $\Omega_M = 0.3$ and $\Omega_\Lambda = 0.7$, with the main matter contribution arising from CDM. The estimates of Ω_Λ from supernovae distances have served to confirm this view. If this turns out to survive further tests, the main challenge for cosmology will be to understand how these values came about.

9.17   Appendix A

In this appendix we give some technical details of the description of the density field in the Universe. If the average density is $\bar{\rho}$, we define the density contrast at any point

$$\delta = \frac{\rho - \bar{\rho}}{\bar{\rho}}.$$

The density contrast δ is a function of position. Its Fourier transform δ_k gives the contribution to the density fluctuations from each wavenumber:

$$\delta(x) = \frac{1}{(2\pi)^3} \int \delta_k e^{-ik\cdot x}\, d^3k,$$

where, if we need to be specific, the spatial coordinates x are comoving coordinates, i.e. physical distances are given by $R(t)x$ and physical wavelengths by $R(t)(2\pi/k)$.

The power spectrum of the density fluctuations is defined as $P_k = \langle|\delta_k|^2\rangle$, where the angle brackets denote an average over a large enough sample.

Note that clustering under gravity means that the density contrasts at different points are not independent. However, while the density contrast remains small, so that we can treat the evolution as a linear perturbation, the Fourier components δ_k are independent.

The mean value of δ is zero. The mean square value is the variance σ^2 given by

$$\sigma^2 = \langle\delta^2\rangle = \frac{1}{2\pi^2} \int_0^\infty \langle|\delta_k|^2\rangle k^2\, dk$$

$$= \int_{-\infty}^\infty \Delta_k\, d(\ln k),$$

where Δ_k is the power per logarithmic interval in wavenumber.

Note that the mass fluctuation on a given mass scale, $\langle\delta M/M\rangle^2$, is also given by σ^2. Ignoring some technical complications, for the mass fluctuations on a scale R we have

$$\left\langle\left(\frac{\delta M}{M}\right)^2\right\rangle \sim \int_0^{1/R} \Delta_k\, d(\ln k).$$

So the mass fluctuations on a given scale R are obtained by integrating over fluctuations on smaller scales. For a power law, $|\delta_k|^2 \propto k^{-\alpha}$ we get a root-mean-square fluctuation in mass on a scale R of $\delta M/M \propto k^3|\delta_k|^2 \propto R^{\alpha+3}$.

The density correlation function is the Fourier transform of the power spectrum of the density contrast:

$$\xi(r) = \langle\delta(x+r)\delta(x)\rangle = \frac{1}{(2\pi)^3} \int \langle|\delta_k|^2\rangle e^{-ik\cdot r}\, d^3k.$$

(As a general relation between autocorrelation functions and power spectra this result is known as the Wiener–Kinchine theorem.) Since the probability of finding a galaxy in a given volume is proportional to the local density we have

$$dP_{12} \propto \langle\rho(x)\rho(x+r)\rangle$$
$$\propto \bar\rho^2\langle(1+\delta(x))(1+\delta(x+r))\rangle$$
$$\propto 1 + \langle\delta(x+r)\delta(x)\rangle$$
$$\propto 1 + \xi(r)$$

from which we recover the definition of the correlation function in the text.

So far we have considered average quantities without regard to the underlying statistical distribution. It is usual to assume that the density contrast is a Gaussian random variable. This means that the real and imaginary parts of each δ_k have a normal distribution. Equivalently, the magnitudes $|\delta_k|$ have a normal distribution with mean zero and variances P_k, and the phases $\arg(\delta_k)$ are randomly distributed with a uniform distribution. It then follows that the statistical properties of the density contrast are determined entirely by the two-point correlation function $\xi(r)$ or, equivalently, the power spectrum P_k. In particular, the higher moments ($\langle \delta(x + r')\delta(x + r)\delta(x)\rangle$ etc) are either zero or related to $\xi(r)$.

9.18   Appendix B

The angular correlation function is defined in terms of the probability of finding galaxy pairs separated by an angle θ on the sky in solid angles $d\omega_1$ and $d\omega_2$. If the mean surface density is $\bar{\sigma}$ per unit solid angle

$$dP_{12} \propto \bar{\sigma}^2 [1 + w(\theta)] \, d\omega_1 \, d\omega_2. \tag{9.14}$$

To establish a relation between $w(\theta)$ and $\xi(r)$ assume that all galaxies have the same absolute magnitude M_*. This means we can assume all galaxies are being counted to the same depth, which simplifies the analysis without affecting the main result. Let this depth be D. The angular probability distribution is obtained from the three-dimensional one by integrating to a depth D:

$$dP_{12} \propto \left\{ \int_0^D r_1^2 \, dr_1 \int_0^D r_2^2 \, dr_2 [1 + \xi(r)] \right\} d\omega_1 \, d\omega_2. \tag{9.15}$$

For an angular separation θ, r is given by

$$r^2 = r_1^2 + r_2^2 - 2r_1 r_2 \cos\theta.$$

Comparing (9.14) and (9.15) we find

$$w(\theta) \propto \int_0^D dr_1 \int_0^D dr_2 \, r_1^2 r_2^2 \xi[(r_1^2 + r_2^2 - 2r_1 r_2 \cos\theta)^{1/2}],$$

which is the required relation. To extract information from it we make two approximations. First, only galaxies at small angular separations are significantly correlated; this enables us to put $\cos\theta \cong 1 - \frac{1}{2}\theta^2$ for small θ. Second, ξ differs from zero significantly for small r only, hence for $r_1 - r_2 \ll r_1 + r_2$. From this it follows that

$$4r_1 r_2 = (r_1 + r_2)^2 - (r_1 - r_2)^2 \approx (r_1 + r_2)^2.$$

It also follows that we can set the limits of integration to be infinite without serious error. Put

$$u = \frac{1}{2D}(r_1 + r_2)$$

$$v = \frac{1}{D\theta}(r_1 - r_2).$$

Then

$$w(\theta) \propto \int_0^\infty du \int_0^\infty dv\, u^4 \xi[(u^2 + v^2)^{1/2} D\theta]\theta$$

gives us the relation between $w(\theta)$ and $\xi(r)$.

Now we come to the main point. Consider a power law correlation function

$$\xi(r) = Ar^{-\alpha}.$$

Then

$$w(\theta) \propto \left\{ \int \int du\, dv\, u^4 (u^2 + v^2)^{-\alpha/2} D^{-\alpha} \right\} \theta^{-\alpha+1}.$$

We conclude that a power-law three-dimensional correlation function gives rise to a power-law angular correlation function. In particular, if the true clustering has no angular scale, then the angular correlation exhibits none either. Usually one assumes the converse also: observation of a power law angular correlation implies a true power-law correlation with no preferred scale of clustering.

9.19   Problems

Problem 74. *Solve equation (9.9) in the general case $R = t^n$. Show that perturbations are frozen-in once the Universe becomes curvature dominated (i.e. once matter density is negligible compared to the curvature term in the Friedmann equation).*

Problem 75. *Derive equation (9.9).*

Problem 76. *Show that for a sphere of radius $\lambda_J/2$ the free-fall collapse time is of order the hydrodynamical timescale λ/c_s and the gravitational energy equals the thermal energy.*

Problem 77. *(a) Show that the mass inside the Hubble sphere M_H grows as $R^{3/2}$ in a matter-dominated model. Find the corresponding rate of growth in the radiation-dominated case.*

(b) Perturbation $\delta\rho/\rho$ grow proportional to R in the matter-dominated case and proportional to R^2 in the radiation-dominated case. For perturbations

$(\delta\rho/\rho)_t$ *outside the horizon at time t show that in both radiation-dominated and matter-dominated models*

$$\left(\frac{\delta\rho}{\rho}\right)_t = M_H^{-2/3} \left(\frac{\delta\rho}{\rho}\right)_H .$$

(See Kolb and Turner 1990, p 365.)

Problem 78. *At early times the curvature of space becomes large. Show that nevertheless at early times the horizon scale is much smaller than the radius of curvature h. Thus the curvature is insignificant within the horizon and Newtonian mechanics is applicable on this scale.*

Problem 79. *Observations of clusters gives a total matter density $\Omega_M \simeq 0.3$. At the same time observation of the microwave background indicates a flat Universe $\Omega_0 = 1$. There must, therefore, be another source of mass energy that does not cluster to make up the difference, the dark energy $\Omega_d(now) \simeq 0.7$. This mass density cannot have been dominant in the past otherwise it would have prevented the formation of structure, so it must grow with time relative to the matter density. Use local energy conservation and the equation of state $P = w\rho c^2$ to show that, at time t,*

$$\rho_d R^{3(\beta+1)} = \text{constant},$$

where $\beta = P/(\rho c^2)$. By considering the ratio Ω_d/Ω_m deduce that the pressure of the dark energy must be negative. (This makes the vacuum energy or quintessence a candidate for the dark energy.)

Problem 80. *Show that the free-streaming length scale for relativistic particles in a radiation-dominated model between time $t = 0$ and $t = t_{nr}$ when the particles become non-relativistic is of order,*

$$\lambda_{fs} \sim 2\frac{t_{nr}}{R_{nr}},$$

where $R_{nr} = R(t_{nr})$. Deduce that $\lambda_{fs} \sim 10(\Omega_p h^2)^{-1}(T_p/T)^4$ Mpc where Ω_p is the contribution of this particle species to the density and T_p the temperature of the species (which differs from the radiation temperature T because the particles are decoupled; see Kolb and Turner 1990 p 352).

Chapter 10

Epilogue

In this chapter we take up some unfinished business. The modern era in cosmology can be divided into two parts. First the isotropic Universe, in which the discovery of the high degree of isotropy confirmed the applicability of the expanding big-bang models. Second, the present anisotropic era, in which it is precisely the departures from isotropy that are the relics from which we hope to reconstruct the details of our past. But anisotropy also raises some general issues to which we turn.

10.1 Homogeneous anisotropy

The Universe is approximately homogeneous and isotropic, but not exactly so. We have seen how we can introduce small inhomogeneities into the homogeneous FLRW models. It is also of interest to see what happens if we introduce anisotropy instead. We can study the exact behaviour of spacetimes with arbitrarily large amounts of anisotropy provided they are homogeneous, principally in order to understand why this situation does not occur. Unfortuneately we have only approximate descriptions of the large departures from homogeneity which we do observe.

In homogeneous anisotropic models at any point the Universe is expanding at different rates in different directions. To visualize this take a small sphere of test particles moving freely in the model Universe. This small sphere of test particles at one instant is distorted into an ellipsoid at other times. This type of motion is referred to as shear. If the axes of the ellipsoid change with time, then we have also a rotational motion. While there are only three ways in which a space can be rotationally symmetric at each point (the open, flat and closed models) the description of anisotropic homogeneous geometries is more complicated. (The conventional scheme is called the Bianchi–Behr classification.) Just as for isotropic models, the simplest known cases are named. As an example we consider the Kasner solution.

10.1.1   Kasner solution

The simplest example of a spatially homogeneous anisotropic space time has the metric form

$$ds^2 = c^2 \, dt^2 - X_1^2(t) \, dx_1^2 - X_2^2(t) \, dx_2^2 - X_3^2(t) \, dx_3^2,$$

which is an obvious generalization of the flat ($k = 0$) FLRW solution with the scale factors $X_1(t)$, $X_2(t)$ and $X_3(t)$ governing the different expansion rates in three orthogonal directions. The mean expansion rate is

$$\frac{\dot{R}}{R} = \frac{1}{3}\left(\frac{\dot{X}_1}{X_1} + \frac{\dot{X}_2}{X_2} + \frac{\dot{X}_3}{X_3}\right), \tag{10.1}$$

the volume expansion is $\theta = 3\dot{R}/R$, the components of shear are

$$\sigma_i = \frac{\dot{X}_i}{X_i} - \frac{\dot{R}}{R}$$

and the rotation is zero. Put $\sigma^2 = \frac{1}{2}(\sigma_1^2 + \sigma_2^2 + \sigma_3^2)$. Then the analogue of the Friedmann equation (5.25) is

$$\tfrac{1}{3}\theta^2 = \sigma^2 + 8\pi G\rho, \tag{10.2}$$

with additional equations for the evolution of shear and expansion also coming from the Einstein equations. The shear acts like an additional energy density in this equation. It evolves as

$$\sigma^2 \propto R^{-6}.$$

As usual, to proceed further we need an equation of state. If we put $p = 0$, then conservation of mass gives $\rho \propto R^{-3}$. Thus, as $R \to 0$, the shear energy dominates the matter term in (10.2) and the spacetime approximates to a vacuum. In fact, if we put $\rho = 0$ exactly, we obtain the vacuum solution

$$X_i \propto t^{p_i}$$

provided

$$p_1 + p_2 + p_3 = 1 = p_1^2 + p_2^2 + p_3^2$$

in order to satisfy Einstein's equations. This is the Kasner solution.

We get $R(t) \propto t^{1/3}$, using the definition (10.1) and the spacetime is singular at $t = 0$ as in the FLRW cases. However, several types of singularity are possible, depending on the relative rates of expansion in different directions. If all of X_1, X_2 and $X_3 \to 0$ as $R \to 0$ we have a 'point' singularity, similar to the isotropic models. If X_1 and $X_2 \to 0$, but X_3 remains finite, we have a 'barrel' singularity with the x_3 axis as the axis of the barrel. If X_1 and $X_2 \to 0$ and $X_3 \to \infty$ we have a 'cigar' singularity. Finally, if $X_1 \to 0$ and X_2 and X_3 remain finite we have a 'pancake' singularity at $t = 0$.

One point to take away from this is just how special in their behaviour the FLRW models are: they are *not* approximations to the general case. This also raises the question of whether other anisotropic models behave similarly. They do not. In particular the generic singularities can be of a much more complicated nature with oscillations in the axes of shear. This led to the idea that these oscillations could be responsible for smoothing the shear in a typical Universe, an idea that has now been overtaken by the inflationary picture.

Of equal interest is the behaviour of the shear as $t \rightarrow \infty$. We have $\sigma/\theta \rightarrow 0$ as $t \rightarrow \infty$, so the shear becomes dynamically unimportant and the expansion approximates increasingly closely to the FLRW behaviour. In addition the cumulative distortions in the microwave background are governed by the integral of the shear $\int \sigma_i \, dt$ back to the last-scattering surface. This integral is finite, guaranteeing that the distortions are small and the model looks almost isotropic to an observer at late times. This does not, however, ensure that the anisotropy in any given model will be less than that observed. Furthermore, unlike the previous simple models, the general anisotropic spacetime contains a mode of shear which starts small and grows in time. This would lead to large distortions of the cosmic background at late times.

10.2 Growing modes

The most direct way of determining the degree of anisotropy in the Universe is through the cosmic background radiation. In the simplest case the residual dipole limits the current shear to $\sigma/\theta < 10^{-4}$. In principle, further constraints on the type of anisotropic model can be obtained from the higher moments.

However, the tightest constraints come from the helium abundance. Equation (10.2) shows that presence of shear speeds up evolution. At the time of nucleosynthesis the constraint is weak, $\sigma/\theta \lesssim 0.5$, but this translates to a current values that can better the limit from the cosmic background by six orders of magnitude.

In view of the anisotropy of the general model Universe that we discussed previously, how is it that the actual Universe is so isotropic? The current view seems to be that inflation comes to the rescue by diluting any reasonable amount of initial anisotropy. But this still leaves us with the problem of the growing modes. Just as inhomogeneity can grow from small beginnings, as we saw in chapter 9, so too can shear (see also Raine and Thomas 1982). And in fact, in the general case, this is precisely what happens, so that at some possibly remote time in the future the Universe will be highly anisotropic. What is it that ensures that the growth of shear will occur in the future and has not occurred by now? One line of argument has it that a period of isotropic expansion is necessary for our existence, perhaps because galaxies cannot grow in the presence of large shearing motions. This is an instance of what has come to be known as the anthropic principle, which says roughly that we cannot observe a Universe in which we do not exist. If there are many Universes or, perhaps more reasonably, many inflationary patches, each having different properties, the anthropic principle would 'explain' some of the properties of our patch, in the same fashion that the conditions for the evolution of life 'explain' the apparently rather exceptional planetary system that we inhabit. This is, of course, true but does not rule out the possibility of a proper explanation.

10.3   The rotating Universe

Constraints similar to those for shear can be obtained for the amount of rotation in the Universe. The crudest estimate would be a rotational velocity less than the measured Doppler shift of 600 km s^{-1} at the Hubble distance, or about one revolution per 10^{13} years. But this can be bettered, by detailed considerations of the microwave anisotropy, by several orders of magnitude.

The rotation of the Universe is of historical interest because it provided Einstein with what he called Mach's principle which was a seminal influence in the development of relativistic cosmology. Mach had argued that the agreement between the Newtonian inertial frames of reference and reference frames fixed (i.e. non-rotating) relative to the stars could not be an accident, but must indicate that the distant matter in the Universe is responsible for determining the inertial frames in some physical way. Einstein's intention in developing relativistic gravity was, in effect, to express the idea that the physical means by which the stars determined inertial frames was through the gravity exerted by distant matter. Unfortunately, Einstein's equations allow cosmologies in which there is inertia but no gravity (because there is no matter) so the equations fail completely in this regard. Worse still perhaps, the equations allow for the existence of Universes that contain matter but in which the local inertial frames rotate relative to the distant stars. The first such example, proposed by Gödel, had certain non-physical features, but has been followed by other more realistic examples. Thus, general relativity fails to provide a basis for Mach's principle in the way that Einstein had intended.

The current view seems to be that inflation will again provide the solution by diluting any physically reasonable initial rotation during the period of inflationary expansion.

10.4   The arrow of time

The Universe appears to evolve from a past to a future. We observe this arrow of time despite the fact that all the physical laws are reversible in time. For any system that evolves towards equilibrium there is a physically allowed system that evolves in the opposite manner in which all the particle motions have been reversed. The former can be observed everywhere, the latter not at all, since we never see systems evolving spontanously away from equilibrium. However, reversing all motions even in a system in equilibrium produces a system in which the velocities are now arranged in a very specific way—the smallest change and the evolution will be completely different from the time-reversal of the approach to equilibrium. We say that the velocities are correlated. The generally agreed reason why we do not observe evolution away from equilibrium is that in an open system (one that can interact with its environment) the evolution destroys correlations. In other words, the environment produces, at the very least, the tiny perturbations to the time-reversed evolution to cause the system to relax back to equilibrium. Thus the time-reverse of a final equilibrium state does not evolve at all, but remains just as good an equilibrium state. Only if we have a closed system, like the Universe as a whole, does this not solve the problem. In this case the correlations are present somewhere and a time-reversed Lazarus would indeed take up his bed and walk (or lie down and die, depending on your point of view).

Various attempts have been made to attach the thermodynamic arrow of time, whereby the future lies in the direction of thermal equilibrium, to the cosmological one. The idea is that the expansion of the Universe provides a sink for correlations. This would imply that a contracting Universe is thermodynamically inconsistent, and indeed various attempts have been made to show that a Universe with above critical density always appears to be expanding thermodynamically. These arguments are not generally accepted. Hence the link between the thermodynamic and cosmological arrows of time is tenuous.

Nevertheless, we believe that low-entropy initial states can evolve to high-entropy final states and not vice versa, and the Universe obliges. The entropy per baryon now is about $10^8 k$, or about $10^{88} k$ per visible Universe. If the mass of the Universe M were collapsed into a black hole its entropy would be $k(M/m_{pl})^2/4$ or something like $10^{120} k$. So this is the available disorder that governs the fate of the world. The operation of gravity can be delayed but it is nonetheless inexorable. Gravity builds stars which evolve to black holes. Black holes grow by swallowing matter and decay by the quantum emission of radiation. The ever-expanding Universe ends in the conversion of matter to an infinite sea of radiation at zero temperature, an infinite sea of useless energy.

And in this process, the operation of gravity appears to produce little islands of sufficient negative entropy in which the Universe can apparently, for a while, be understood. But the evolution of structure demands an arrow of time, and that arrow points to the dissolution of structure into a featureless state of maximum entropy. The Universe, it would appear, evolves through just that state in which it can know its own oblivion. Throughout all the galaxies, on countless shores of fragile green, countless intelligences discover the Universe to be merely a joke. This, it seems, is the vision written in the patterns of all those stars, from which we fashion a Universe amidst its black amnesias.

Reference material

Constants

Elementary charge	e	1.602×10^{-19} C
Electron rest mass	m_e	9.109×10^{-31} kg
Proton rest mass	m_p	1.673×10^{-27} kg
Neutron rest mass	m_n	1.675×10^{-27} kg
Planck constant	$\hbar$	$(6.626/2\pi) \times 10^{-34} = 1.0546 \times 10^{-34}$ J s
Speed of light in vacuum	c	2.998×10^{8} m s^{-1}
Gravitational constant	G	6.673×10^{-11} N m^2 kg^{-2}
Stefan constant	σ	5.671×10^{-8} W m^{-2} K^{-4}
Radiation constant	$a = 4\sigma/c$	7.564×10^{-16} J m^{-3} K^{-4}
Avogadro number	N_A	6.022×10^{23} mole^{-1}
Thomson cross section	σ_T	6.652×10^{-29} m^2
Boltzmann constant	k	1.381×10^{-23} J K^{-1}
Permittivity of vacuum	ε_0	8.854×10^{-12} F m^{-1}
Permeability of vacuum	μ_0	$4\pi \times 10^{-7}$ H m^{-1}
Atomic mass unit	u	1.661×10^{-27} kg $= 931$ MeV
1 eV		1.602×10^{-19} J
1 keV		1.161×10^{7} K
1 parsec		3.086×10^{16} m
Seconds in a year		3.156×10^{7} s year^{-1}
Solar mass	$M_\odot$	1.989×10^{30} kg
Solar luminosity	$L_\odot$	3.846×10^{26} W

Useful quantities

Microwave background temperature	T_0	2.725 ± 0.001 K
Number density of background photons	n_γ	$4.105 \times 10^{8} (\frac{T}{2.725})^3$ m^{-3}
Energy density of background photons	$u_\gamma = aT^4$	$4.170 \times 10^{-14} (\frac{T}{2.725})^4$ J m^{-3}
Number density of a $\nu\bar{\nu}$ family	$n_{\nu\bar{\nu}}$	$\frac{3}{11} n_\gamma$
Hubble constant	H_0	$100h$ km s^{-1} Mpc^{-1}
Hubble time	H_0^{-1}	$9.779 \times 10^{9} h^{-1}$ year
Critical density	$\rho_c = \frac{3H^2}{8\pi G}$	$1.88 \times 10^{-26} h^2$ kg m^{-3}

Formulae

Friedmann equation

$$\left(\frac{dR}{dt}\right)^2 = \frac{8}{3}\pi G\rho R^2 - kc^2 + \frac{1}{3}\Lambda R^2.$$

Energy equation

$$\frac{d(\rho R^3)}{dt} + \frac{p}{c^2}\frac{dR^3}{dt} = 0.$$

Acceleration equation

$$\frac{d^2 R}{dt^2} = -\frac{4}{3}\pi GR(\rho + 3p/c^2) + \frac{1}{3}\Lambda R.$$

Evolution of Hubble parameter

$$H(z) = H_0[(1+z)^2(1+z\Omega_M) - z(2+z)\Omega_\Lambda]^{1/2}.$$

Redshift–time relation

$$\frac{dz}{dt} = -(1+z)H.$$

For large z:

$$t \sim \tfrac{2}{3}(1+z)^{-3/2}H_0^{-1}\Omega_0^{-1/2}.$$

Temperature–time relation (radiation-dominated):

$$T = 1.6 g_*^{-1/4} t^{-1/2} \text{ MeV}$$
$$= 2 \times 10^{10} g_*^{-1/4} t^{-1/2} \text{ K}.$$

Einstein–de Sitter model:

$$\frac{R(t)}{R_0} = (6\pi G\rho_0)^{1/3} t^{2/3}$$
$$\simeq 3 \times 10^{-12} h^{2/3} t^{2/3},$$

(t in seconds).

Definition of (astronomical) apparent magnitude difference of sources with luminosities L_1 and L_2:

$$m_1 - m_2 = 2.5 \log(L_2/L_1).$$

The absolute magnitude of a source at distance d pc is

$$m - M = 5 \log d - 5.$$

To set the zero point we can use the fact that the absolute magnitude of the Sun in the visual waveband is $M_{v\odot} = 4.79$ and its known distance and luminosity.

Symbols

ρ_b	baryon mass density at a general time
ρ_γ	equivalent mass density in photons at a general time
ρ_λ	equivalent vacuum mass density (corresponding to a cosmological constant) at a general time
ρ_m	total mass density in baryons and dark matter at a general time
ρ_r	equivalent mass density in radiation, including both photons and neutrinos, at a general time
ρ_M, ρ_R, ρ_B, ρ_Λ	are the corresponding quantities at the present time
ρ_0	the total mass density at the present time
Ω_i	are the corresponding density parameters defined by $\Omega_i = 4\pi G \rho_i/(3H^2)$
s_i	are the corresponding entropy densities
u_i	are the corresponding energy densities
$R(t)$	the scale factor at time t
R_0	the scale factor at the present time t_0
g_*	the effective number of degrees of freedom of all relativisitic particles at a given time
η^{-1}	the number of photons per baryon subsequent to $e^{\pm}$ annihilation
η_{tot}^{-1}	the total entropy per baryon (including all relativistic species)
H	the Hubble parameter (at any time)
H_0	the Hubble constant (H at the present time)
q	the deceleration parameter at any time
q_0	the current value of the deceleration parameter

References

Chapter 1

Pagel B J E 1997 *Mon. Not. R. Astron. Soc.* **179** 81P

Chapter 2

Boyle B J, Shanks T, Georgantopoulos I, Stewart G C and Griffiths R E 1994 *Mon. Not. R. Astron. Soc.* **271** 639
Dressler A *et al* 1987 *Ap. J.* **313** 42
Harrison E R 2000 *Cosmology: the Science of the Universe* 2nd edn (Cambridge: Cambridge University Press) p 274
Hubble E 1929 *Proc. Natl Acad. Sci.* **15** 168
Sandage A 1988 *Ann. Rev. Astron. Astrophys.* **26** 561–630
Turner M S and Tyson J A 1999 *Rev. Mod. Phys.* **71** S145
Wall J V 1994 *Aust. J Phys.* **47** 625

Chapter 3

Ahmad Q R *et al* 2001 *Phys. Rev. Lett.* (in press)
Bahcall N A, Lubin L and Dorman V 1995 *Ap. J.* **447** L81
Balbi A *et al* 2000 *Ap. J.* **545** L1
Evrard A E 1997 *Mon. Not. R. Astron. Soc.* **292** 289
Fukugita M *et al* 1998 *Ap. J.* **503** 518
Gates E I *et al* 1995 *Phys. Rev. Lett.* **74** 3724
Hagmann C *et al* 1998 *Phys. Rev. Lett.* **80** 2043
Jungman G *et al* 1996 *Phys. Rep.* **267** 195
Milgrom M 1986 *Ap. J.* **306** 9
Mulchaey J S *et al* 1996 *Ap. J.* **456** 80
Peebles P J E 1993 *Principles of Cosmology* (Princeton, NJ: Princeton University Press)
Press W H and Spergel D N 1985 *Astrophy. J.* **296** 679
Sadoulet B 1999 *Rev. Mod. Phys.* **71** S197
Schechter P 1976 *Ap. J.* **203** 297
Schramm D and Turner M 1998 *Rev. Mod. Phys.* **70** 303
Steigman G 1976 *Ann. Rev. Astron. Astr.* **14** 339
Tripp T M *et al* 2000 *Ap. J.* **534** L1
Turner M S 1999 *Phil. Trans. R. Soc.* A **99** 357

Turner M S and Tyson J A 1999 *Rev. Mod. Phys.* **71** S145
Tytler D *et al* 1996 *Nature* **381** 207
von Albada T S and Sancisi R 1986 *Phil. Trans. R. Soc.* A **320** 447
White S D M *et al* 1993 *Nature* **366** 429
Zaritzky D *et al* 1993 *Ap. J.* **405** 464

Chapter 4

Comestri A *et al* 1995 *Astron. Astrophys.* **296** 1
Dicke R H *et al* 1965 *Ap. J.* **142** 414
Ellis G F R 1971 General relativity and cosmology *Proc. Int. School of Physics, E Fermi, Course XLVII* ed R K Sachs (New York: Academic) p 104–82
Fixsen D J *et al* 1996 *Ap. J.* **473** 576
Harrison E R 1995 *Ap. J.* **446** 63
Hasinger G *et al* 2000 *Nature* **404** 443
Hauser M G *et al* 1998 *Ap. J.* **508** 25
Kraan-Korteweg R C *et al* 1996 *Nature* **379** 519
Mather J C *et al* 1990 *Ap. J.* **354** L37
——1999 *Ap. J.* **512** 511
Miyaji T and Boldt E A 1990 *Ap. J.* **253** L3
Muller R A 1978 *Sci. Am.* **238** 64
Partridge R B 1995 *The Cosmic Background Radiation* (Cambridge: Cambridge University Press)
Peacock J A 1999 *Cosmological Physics* (Cambridge: Cambridge University Press) p 357
Peebles P J E 1993 *Principles of Physical Cosmology* (Princeton, NJ: Princeton University Press) p 139
Penzias A A and Wilson R W 1965 *Ap. J.* **142** 419
Puget J-L *et al* 1996 *Astron. Astrophys.* **308** L5
Rybicki G B and Lightman A P 1979 *Radiative Processes in Astrophysics* (New York: Wiley) ch 4
Smoot G F *et al* 1977 *Phys. Rev. Lett.* **39** 898
——1992 *Ap. J.* **396** L1
Songaila A *et al* 1994 *Nature* **371** 43
Srianand R, Petitjean P and Ledoux C 2000 *Nature* **408** 931
Treyer M *et al* 1998 *Ap. J.* **509** 531
Tyson J A 1995 *Extragalactic Background Radiation (STSI Symposium Series 7)* ed D Calzetti *et al* (Cambridge: Cambridge University Press) p 103
Weinberg S 1977 *The First Three Minutes* (London: Bantam Books) ch 3
Wright E L *et al* 1994 *Ap. J.* **420** 450
Wu K K S *et al* 1999 *Nature* **397** 225

Chapter 5

d'Inverno R 1992 *Introducing Einstein's Relativity* (Oxford: Oxford University Press)
Harrison E R 1991 *Ap. J.* **383** 60
Perlmutter S *et al* 1998 *Nature* **391** 51
——1999 *Ap. J.* **517** 565

Will C M 1993 *Theory and Experiment in Gravitational Physics* (Cambridge: Cambridge University Press)

Chapter 6

Bahcall S R and Tremaine S 1988 *Ap. J.* **326** L1
Chaboyer B *et al* 1998 *Ap. J.* **494** 96
Chiba M and Yoshii Y 1999 *Ap. J.* **510** 42
Dabrowski Y *et al* 1997 *Mon. Not. R. Astron. Soc.* **277** 753
Ellis R S 1997 *Ann. Rev. Astron. Astrophys.* **35** 404
de Bernardis P *et al* 2000 *Nature* **404** 955
Filippenko A V 1997 *Ann. Rev. Astron. Astrophys.* **35** 309
Goldhaber G and Perlmutter S 1998 *Phys. Rep.* **307** 325
Guerra E J and Daly R A 1998 *Ap. J.* **493** 536
Jones M A and Fry J N 1998 *Ap. J.* **500** L75
Kellermann K I 1993 *Nature* **361** 134
Kochaneck C S 1996 *Ap. J.* **494** 47
Kragh H 1996 *Cosmology and Controversy* (Princeton, NJ: Princeton University Press)
Liebundgut B *et al* 1996 *Ap. J.* **466** L21
Linweaver C H 1999 *Science* **284** 1503
Loh E D and Spillar E J 1986 *Ap. J.* **307** L1
Phillips M M 1993 *Ap. J.* **413** L105
Perlmutter S *et al* 1998 *Nature* **391** 51
——1999 *Ap. J.* **517** 565
Riess A G *et al* 1997 *Astron. J.* **114** 722
——1998 *Astron. J.* **116** 1009
Sandage A 1988 *Ann. Rev. Astron. Astrophys.* **26** 561
Tipler F, Clarke C and Ellis G 1980 *General Relativity and Gravitation* vol 2, ed A Held (New York: Plenum) pp 97–206
Zeilik M, Gregory S A and Smith E V 1997 *Introductory Astronomy and Astrophysics* (New York: Wiley)

Chapter 7

Allen C W 1973 *Astrophysical Quantities* (London: Athlone)
Burles S *et al* 1999 *Phys. Rev. Lett.* **82** 4176
Kolb E W and Turner M S 1990 *The Early Universe* (Reading, MA: Addison-Wesley)
Liebundgut B *et al* 1996 *Ap. J.* **466** L21
Padmanabhan 1993 *Structure Formation in the Universe* (Cambridge: Cambridge University Press)
Peebles P J E 1971 *Physical Cosmology* (Princeton, NJ: Princeton University Press)
Schramm D N and Turner M S 1998 *Rev. Mod. Phys.* **70** 303
Schwarzschild B 2000 *Phys. Today* **53**(5) 20
Shiozawa M *et al* 1998 *Phys. Rev. Lett.* **81** 3319
Wagoner R V 1973 *Ap. J.* **179** 343

Chapter 8

Ellis G F R and Rothman T 1993 *Am. J. Phys.* **61** 883
Guth A H 1981 *Phys. Rev.* D **23** 347
Harrison E R 1991 *Ap. J.* **383** 60
——2000 *Cosmology: the Science of the Universe* 2nd edn (Cambridge: Cambridge University Press)
Kolb E W and Turner M S 1990 *The Early Universe* (Reading, MA: Addison-Wesley)
Tryon E P 1973 *Nature* **246** 396

Chapter 9

Abell G O 1958 *Astrophys J. Suppl.* **3** 211
Arfken G B and Weber H J 1995 *Mathematical Methods for Physicists* 4th edn (New York: Academic)
Bahcall N A 1988 *Ann. Rev. Astron. Astrophys.* **26** 631–86
de Bernardis P *et al* 2000 *Nature* **404** 955
de Vaucouleurs G and de Vaucouleurs A 1964 *Reference Catalogue of Bright Galaxies* (Houston, TX: University of Texas Press)
Kaiser N 1984 *Ap. J.* **284** L9
Kolb E W and Turner M S 1990 *The Early Universe* (Reading, MA: Addison-Wesley)
Margon B 1999 *Phil. Trans. R. Soc.* A **357** 93–103
Peebles P J E 1971 *Physical Cosmology* (Princeton, NJ: Princeton University Press)
Schectman S A 1996 *Ap. J.* **470** 172
Shane C D and Wirtanen C A 1967 *Publ. Lick Observatory* **22** 1
Shanks T *et al* 1989 *Mon. Not. R. Astron. Soc.* **237** 589
Shapley H and Ames A 1932 *Ann. Harvard College Observatory* **88** 43
Wu K K S, Lahav O and Rees M J 1999 *Nature* **397** 225

Chapter 10

Raine D J and Thomas E G 1982 *Astrophys. Lett.* **23** 37

Index

absorption timescale, 149
acceleration equation, 75
adiabatic fluctuations, 196
adiabatic perturbations, 192, 193, 197
age, 83, 85, 86, 88, 118, 119
age problem, 122
angular correlation function, 202
angular diameter, 104
anthropic principle, 207
anti-helium, 38
apparent magnitude, 107
axions, 34, 35

B-violation, 139
baryogenesis, 138, 174
baryonic matter, 30, 33
β-decay, 144
bias, 187
binding energy, 144
blackbody radiation, 156
blue shift, 10
bolometric luminosity, 108
Boltzmann formula, 144
BOOMERanG, 199, 200
bosons, 130
bottom-up galaxy formation, 180
bremsstrahlung, 48, 138
bubble collisions, 172

carbon, 146
Cepheid variables, 14
chaotic inflation, 176
chemical potential, 130, 144, 154
closed space, 70

cluster correlation function, 187
cluster mass, 30
COBE satellite, 43, 45, 54, 57, 198
cold dark matter, 180, 197, 198
Coleman–Weinberg potential, 175
Coma cluster, 32
comoving observers, 70
comoving volume, 24
Compton scattering, 48, 138, 150
Copernican principle, 11
correlation function, 183, 202
cosmic background radiation, 18, 45, 46
cosmic strings, 176, 194
cosmic time, 71
cosmic-rays, 38, 42
cosmological constant, 75, 87, 113
cosmological principle, 11, 13, 17, 46, 58, 65, 71
    perfect, 48
CP symmetry, 35
CP violation, 139
critical density, 21, 77, 88, 121
curvature perturbations, 192
curved spacetime, 64, 65

dark energy, 97
dark matter, 24, 27, 30, 31, 34–36, 180, 197
    CDMS collaboration, 37
    DAMA, 37
deceleration parameter, 16
degrees of freedom, 131, 212
density contrast, 200

density correlation function, 201
density parameter, 21, 78–80, 142
deuterium, 31, 145
dipole anisotropy, 47, 53
dipole fluctuation, 199
distance ladder, 14
domain walls, 176
dominant energy condition, 97
Doppler effect, 11, 73
Doppler peak, 199
Doppler shift, 10, 15, 26, 49, 199
double Compton scattering, 138
dust model, 78

Einstein model, 98
Einstein–de Sitter, 80, 82, 92
electron–positron annihilation, 140
electron–positron pairs, 158
element abundances, 146
elliptical galaxies, 28
energy
    conservation, 53
    conservation equation, 74
energy density, 51, 131
entropy, 52, 140, 141, 209
    density, 52
    per baryon, 52, 58
entropy conservation, 140
entropy density, 141, 158
entropy per baryon, 159
equation of state, 75, 78, 90, 96,
    169
Euclidean geometry, 61, 65, 69
expansion, 11
expansion timescale, 137

faint blue galaxies, 117
false vacuum, 168, 173, 176
Fermi–Dirac spectrum, 130
fermions, 130
field equations, 64, 74, 75
flatness problem, 121, 165
FLRW models, 61, 75
fluctuation spectrum, 180, 194

flux, 108
free fall, 62
free–free absorption, 149
free-streaming, 197
Friedmann, 60
Friedmann equation, 74, 79
fundamental observers, 70

galactic magnetic field, 41, 42
Galaxy, 55
galaxy catalogues, 181
galaxy clusters, 13, 23, 29
galaxy correlation function, 186
galaxy distribution, 183
galaxy formation, 165, 176
gamma-ray background, 56
Gamov criterion, 137
general relativity, 62
globular clusters, 118, 119
gravitational lenses, 32
gravitational lensing, 119
gravitational waves, 44
Great Attractor, 55
Gunn–Peterson test, 33
GUT era, 138
Guth, 173
GUTs, 175

Harrison–Zeldovich spectrum, 195,
    196
helium, 118, 130, 145, 146, 207
Hertzprung–Russel diagram, 118
Higgs field, 175, 176
Hipparcos, 14
homogeneity, 67
homogeneous anisotropy, 205
horizon distance, 120
horizon problem, 120, 164
hot dark matter, 180, 197
Hubble, 14, 60
    space telescope, 14
Hubble constant, 16
Hubble flow, 14
Hubble parameter, 15, 80

Hubble plot, 107, 111
Hubble's law, 14, 74
Hubble sphere, 93, 100, 166
Hubble telescope, 33
hyperbolic geometry, 69

inflation, 163, 166, 170
inflaton, 168, 172
infrared background, 43
intracluster gas, 30, 31, 57
isocurvature perturbations, 192, 193
isothermal fluctuations, 196
isothermal perturbations, 193
isotropy, 11, 45, 67

Jeans' mass, 191–193

K-correction, 110
Kasner solution, 206
kinetic equilibrium, 138

Las Campanas, 182
last scattering, 153
latent heat, 173
Lemaître, 60
Lemaître redshift relation, 50
Lemaître redshift rule, 71
lepton number, 143
light cones, 93, 95
light elements, 146
line element, 64
Liouville's theorem, 133
Local Group, 11, 28
luminosity evolution, 117
luminosity function, 25

Mach's principle, 208
magnetic monopoles, 176
main sequence, 118
MAP satellite, 22, 113, 199
mass
    conservation, 78
mass density, 21, 23, 30
mass fluctuations, 201
mass to light, 24, 25, 27, 28, 30

massive neutrinos, 34
matter dominated, 78
matter radiation equality, 90, 135
Mattig relation, 102
MAXIMA, 199, 200
Maxwell–Boltzmann distribution,
    132
metric, 64
Milne model, 98

negative pressure, 76, 97, 169
neutralinos, 34, 36
neutrino background, 34, 44, 129,
    155
neutrinos, 34, 35, 140, 157
neutron freeze-out, 143, 148
neutron lifetime, 144
new inflation, 174
Newtonian interpretation, 81
nucleosynthesis, 30, 33, 142
number counts, 113

Olber's paradox, 18
open space, 70
optical background, 43

particle horizon, 95
perfect gas, 130
phase transition, 167, 172, 174
photon number, 142, 157
photon-to-baryon ratio, 146
Planck satellite, 22, 113, 199
Planck spectrum, 50, 51, 130
Poisson's equation, 188
power spectrum, 198, 201
principle of equivalence, 62
proper distance, 69, 93
proper time, 62, 63
proper volume, 70

quadrupole fluctuation, 199
quantum fluctuations, 167, 172
quantum tunnelling, 172, 175
quintessence, 97

radiation dominated, 90, 136, 166
radiation pressure, 52
radiation temperature, 128
radiation Universe, 90
radio background, 41
radio galaxies, 106
random distribution, 184
random walk, 149
recession velocity, 12
recombination, 151
redshift, 10
redshift surveys, 181
reheat temperature, 174
reheating, 173
relativistic matter, 91
rest frame, 55
Robertson–Walker metric, 66, 68,
    70, 71, 73, 74
Robertson–Walker models, 61
rotation, 208
rotation curve, 26

Saha equation, 145, 152, 159
scale factor, 12, 15, 65
Schecter function, 25
shear, 205, 207
Silk mass, 194
singularities, 206
singularity problem, 122
Sloan survey, 182
slow-roll, 171, 176
spacetime diagram, 92
special relativity, 62
spherical geometry, 69
spiral galaxies, 26, 27
standard candle, 14, 107, 110
static Universe, 60
steady-state theory, 48
stiff matter, 97, 169
strong energy condition, 123

superclusters, 182, 193
SuperKamiokande, 38
supernovae, 14, 30, 111
supersymmetry, 36
symmetry breaking, 167, 175, 176

temperature–time relation, 91
thermal equilibrium, 48, 138, 154
thermalization, 138, 174
Thomson cross section, 140
Thomson scattering, 148
tilt, 172
time dilation, 74, 124
timescale test, 118
tired light, 124
top-down galaxy formation, 180,
    193
2DF survey, 182
two-point correlation function, 184

vacuum, 167
vacuum energy, 171
velocity–distance law, 13, 73, 74,
    94
Virgo Supercluster, 14, 38
virial theorem, 24, 28, 39
voids, 182, 197

warm dark matter, 197
weak energy condition, 97
WIMPS, 34, 36
world map, 92
world picture, 92

X-bosons, 138
x-ray background, 44, 56
x-ray dipole, 56
x-rays, 31, 33

Zeldovich–Sunyaev effect, 154